A FOOD FOREST
in your Garden

Plan it — Grow it — Cook it

Alan Carter

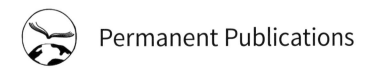
Permanent Publications

Published by
Permanent Publications
Hyden House Ltd
The Sustainability Centre
East Meon
Hampshire
GU32 1HR
United Kingdom
Tel: 01730 776 582
          International: +44 (0)1730 776 582
Email: enquiries@permaculture.co.uk
Web: www.permanentpublications.co.uk

Distributed in North America by
Chelsea Green Publishing Company, PO Box 428, White River Junction, VT 05001, USA
www.chelseagreen.com

Designed by Two Plus George Limited, info@twoplusgeorge.co.uk

Printed in the UK by Bell & Bain, Thornliebank, Glasgow

All paper from FSC certified mixed sources

The Forest Stewardship Council (FSC) is a non-profit international organisation
established to promote the responsible management of the world's forests.
Products carrying the FSC label are independently certified to assure consumers
that they come from forests that are managed to meet the social, economic and
ecological needs of present and future generations.

British Library Cataloguing-in-Publication Data
A catalogue record for this book is available from the British Library

ISBN 978 1 85623 299 9

Praise for the book

Whether you want to follow in Alan's footsteps and become a dedicated forest gardener, or are simply curious about this fascinating and sustainable way of producing food, you will love this book. Having converted his allotment into a haven for wild food he dispels a common myth that the first requirement of a forest garden is a forest!

Steve Ott
editor of *Kitchen Garden* magazine

Alan walks us through the theory and reality of creating a food forest, whatever the size of our plot. As the processes and delights of creating a home garden unfold, we are guided by useful observations and practical information. The directory of forest garden plants is an eye-opener and has me rethinking so many plants I've dismissed in the past. I wish I'd read this book before creating our first food forest!

Liz Zorab
Byther Farm, author of *Grounded*

Alan has written a fantastic comprehensive book I would have loved to have had time to write myself, covering a multitude of edible plants suited to all the diverse habitats that can make up a forest garden, and with particularly relevance to colder climates. I genuinely enjoyed this book and learned plenty from it; not really a surprise as I've enjoyed Alan's blog with the curious name 'Of Plums and Pignuts' and sought him out when I was passing Aberdeen. Importantly, the book provides the inspiration to growing more climate friendly food in what are often called 'marginal areas'. I look forward to making pernicious pasta every year!

Stephen Barstow
grower, speaker and author of
Around the World in 80 Plants

A Food Forest in Your Garden offers a concise yet incredibly comprehensive guide for the many people interested in creating diverse perennial food gardens at home. Alan Carter conveys his extensive knowledge clearly, in a format accessible to all gardeners, both new and well-seasoned!

Caroline Aitken
author and teacher at Whitefield Permaculture

It's great to see a forest gardening book written with Scottish conditions in mind. This is one of the best recent books I've read aimed at smaller-scale forest gardens too, and is especially good with its coverage of the many herbaceous crops it is possible to grow.

Martin Crawford
founder of the Agroforestry Research Trust and
author of *Trees for Gardens, Orchards
and Permaculture*

This is the book we in the North have been waiting for, as Alan Carter has a realistic understanding of the climate. He comes up with a remarkable range of edible plants that will 'do' and all the techniques needed to get them established, making a British forest garden seem wholly desirable and possible.

Fi Martynoga
writer, forager, gardener and editor of
Scotland's Wild Harvests

A brilliant, down-to-earth guide to creating your own small-scale forest garden, based on solid ecological understanding and brimming with hands-on experience.

Tomas Remiarz
teacher, designer and author of
Forest Gardening Practice

Alan Carter has always impressed me as a follower and as a leader. If you enjoy this book it will be because of his very practical knowledge, wisdom achieved over many years, by trial and error and by observation – as all the best learning is done. His deep spiritual commitment to the process is shared by all true Forest Gardeners. We don't just love our gardens, they love us! Beautifully illustrated – every page of this book speaks from the heart. If you need any more incentive than this to enrich your life and that of your family and friends, I don't know where you'll find it! We invest in the wisdom of others that we may ourselves share it onwards.

Graham Bell
permaculture teacher and author of
The Permaculture Garden and *The Permaculture Way*

Neatly contained in Alan's timely book is the knowledge to achieve greater productivity from your plot for many aspiring gardeners. You will view 'untidy' and 'weedy' gardens very differently after studying it, and see the role of tree fruit in a more important light. I thoroughly recommend it!

Andrew Lear
fruit tree grower and nurseryman
at Appletreeman

Events over the last year or so have shown us that along with so much else, the old 'Business as Usual' food production and supply model needs to change. Any 'New Normal' will mean re-localising on every scale, including learning to grow more produce in our own gardens, allotments or whatever spaces we can find, however small. The vision of highly abundant and ecologically regenerative edible landscapes presented in this timely and informative volume both reminds and inspires us that we can all play our part in Growing Back Better.

Graham Burnett
permaculture teacher and author of several books,
including *The Vegan Book of Permaculture*

Alan takes us on a journey that empowers us to plant longer-term food in our gardens and growing spaces. He draws on a diverse range of experiences and puts it down for the reader, so they know what to do. This is a comprehensive guide and a mine of information, sure to be an asset on your bookshelf.

Anna Locke
educator and author of *The Forager's Garden*

Thirty years of research, experimentation and experience, and 10 years of teaching about planning, growing and eating from his forest garden of 200 square metres in Aberdeenshire are laid out here by Alan. From principles to design, through management to eating, with a directory of edible plants and fungi, all well referenced, this book is a goldmine.

John Mclennan
founder of 'A Greener Melrose', keen gardener
and environmental activist

In all my years of designing and implementing forest gardens, I am always seeking out simple, focused, and step-by-step design and implementation manuals. But, first and foremost, after sixty-plus years of working with plants, I have found that management, maintenance, and the transformation of plants into food is key to longevity, the upgrading of fertility on a yearly basis, and an understanding of healthy balance of trees, shrubs, herbaceous species, and most of all, human beings. I cannot recommend Mr. Carter's excellent and succinct approach to forest gardening enough: step-by-incremental step, fine charts and visuals, a much-needed chapter on harvest and cooking, and a thoroughly enjoyable directory of species for forest gardens. Buy it, study it in depth, do it. There is no substitute to putting these systems on the ground and observing their functions, with persistence, through the years. This manual gives you the steps to get there.

Wayne Weiseman
author of *Integrated Forest Gardening*

Acknowledgements

Plants and knowledge both have the magical quality that they are increased, not decreased, by sharing. Forest gardening, perennial vegetable growing and amateur plant breeding are generous and collaborative fields and I am grateful to everyone who has shared their knowledge and seeds with me over the years.

I would like to thank Addie Fern, Michaela Hunter, Keveral Farm, Graham Bell and Nancy Woodhead, and Penny and Martin Sherring for showing me around their forest gardens. Emma Planterose – I'll get there yet!

To the staff and Friends of the Cruickshank Botanic Garden, thank you for seeds from the Index Seminum, permission to take samples for experiments, general interest and encouragement and maintaining such a beautiful space on my doorstep. To the allotment team at Aberdeen City Council, thank you for giving me a space for my experiments and for understanding that I am cultivating my allotment even if it is not in the normal way.

The users of the Facebook discussion groups Plant Breeding for Permaculture; Radix Root Crops; Edimentals and Perennial Vegetables; and the discussion boards at Home Grown Goodness have given me countless tips and much food for thought.

Many thanks to Stephen Barstow for doing so much to both pioneer and generously support the field of growing perennial fruit and vegetables at improbable latitudes. The seeds of my experience with many of the species in this book came first from Stephen, both literally and figuratively.

I'm grateful to Oihane, Steve, Julian and Khemaka for reading and offering comments on the text.

Many people generously allowed me to use their photographs and are credited with the pictures. Thanks in particular to Andy Coventry, Julian Maunder, Tim Hill, Stephen Barstow and the users of Wikimedia Commons. Much appreciation to Grace Banks and Josian Llorente Sesma for help with diagrams and illustrations.

This book was largely written in the unexpected circumstances of lockdown for COVID-19. Thank you to Oihane for love, support and cups of tea, which helped enormously in getting me through the process.

THE AUTHOR

Alan studied forestry and has worked variously in
forestry, gardening, conservation and greenspace
management. He has been writing and teaching
about forest gardening since 2011, having spent
many years experimenting with it in his allotment
in Aberdeen.
www.foodforest.garden

Contents

Part 1. Techniques for a Forest Garden

Part 2. A Directory of Forest Garden Plants

PART 1

TECHNIQUES FOR A FOREST GARDEN

Hop shoot emerging

1

Understanding Forest Gardens

Introduction

Forest gardens are not new. They have been grown in the tropics for centuries, if not millennia. In Indonesia they are called home gardens or *pekarangan*. Like a forest, a home garden contains a great variety of plants in many layers, from the tall trees of the canopy to shrubs and shade-loving plants on the forest floor. The difference is that in the home garden all the plants have been chosen for their use, which might be as food, building materials, fibres or medicines. The canopy might include coconut, durian, jackfruit and timber trees, with bananas, papaya and coffee lower down, and vegetables and starchy crops such as aubergines, chillies, cassava, sweet potatoes and taro in the lowest layers. Animals including fish in ponds are also usually part of the system. Researchers call the cultivation of such edible ecosystems 'agro-ecology'.

The name 'home garden' is not accidental. In contrast to rice fields and shifting cultivation out in the forest, a *pekarangan* is always around the house. The garden is an important part of the household and the family's activities are an important part of the garden. As well as material production, a home garden is the venue for social, cultural and religious activity – to the extent that some researchers have suggested that the term should be extended into 'agro-socio-ecology'.

My own introduction to the idea came when I was studying forestry at the University of Aberdeen in the early 1990s. I saw a poster advertising a talk on 'Indonesian home gardens: sustainable development without aid?' and went along. The speaker was Mike Daw, a UN Food and Agriculture Organisation advisor who travelled around the world studying and giving advice on agricultural systems. One of his trips had taken him to Indonesia. The gist of his talk was that he had found almost nothing that he could say to improve home gardens: they were already a perfect system.

Naturally this stuck with me, since I am both a forester and a keen gardener, but the chances of growing coconuts and bananas in Aberdeen are limited. It was only later that I discovered that attempts were under way to adapt the system to temperate climates. The pioneer in the UK is generally acknowledged to be Robert Hart, who planted a temperate home garden on his smallholding in Wenlock Edge in Shropshire after the Second World War. Hart adopted the term 'forest garden', by which home gardens are usually known in Britain.[*] His garden and his writings inspired a new generation of forest gardeners such as Ken and Addie Fern of Plants For A Future, Martin Crawford, who runs the Agroforestry Research Trust in Devon,

[*] This does sometimes lead to confusion as it sounds like a forest garden might be a garden in a forest. I usually explain that the garden isn't *in* the forest, it *is* the forest. In Scotland at least, 'woodland garden' seems to have emerged as the term for the ornamental variety and 'forest garden' for the edible. For the former, see *Woodland Gardening* by Kenneth Cox.

and Graham Bell and Nancy Woodhead in the Scottish Borders.

As soon as I got my own garden, I also started to experiment with the perennial vegetables and fruits that grow in a forest garden. I soon found that what grows in a favourable microclimate in the southwest of England, where most of the research in the UK has been done, doesn't necessarily do so well in the yet cooler climate of the north of Scotland. I also discovered that there is a difference between 'edible' in the sense of 'you can eat this plant without dying' and edible in the sense that you would want to. Since then I have been working out what will grow in my garden and what I want to bring into my diet and my daily life. My style of forest gardening is also influenced by the relatively small space I have for it: my main growing space is a standard council allotment of around 200 square metres.

One thing I have discovered is that the adaptations I have had to make to a cooler climate and to a smaller garden are similar. As you go north (or higher) in Britain, it is not just that the species change. Forest structure also changes, becoming simpler and less layered. Both these things bring the focus of a productive forest garden down from the trees to the shrubs and ground layer and change the ecological model from closed high forest to the more open forest edge.

My reward for these experiments has been a cool-temperate home garden that feeds both my belly and my heart. This book is my way of passing on what I have learned. I hope it will be of benefit to anyone interested in trying this style of gardening and growing food on any less-than-ideal site or smaller garden in the cool temperate world. Think of it as the north giving something back to a field that has gained so much from the south.

What is a forest garden?

Forest gardening is the way to have your garden and eat it. When I first got my allotment I planted an area with food crops, rotated with green manure crops to feed the soil. I planted another area with attractive flowers and another with 'plants for wildlife' for pollinators and birds. As my forest gardening experiments have expanded, I have discovered more and more that my crop plants fulfil all the aims I had planned for these different areas. They feed me, they feed the wildlife, they build the soil and they lift the heart, all at the same time. My forest garden produces a similar amount of food to an annual garden of the same size, but I get a garden full of wildlife and a beautiful place to be at the same time, for free.

To understand why, consider the difference between a perennial crop plant and a close annual/biennial equivalent. For anyone unfamiliar with these terms: annuals grow, seed and die in one year and biennials in two; perennials carry on for years. Perennials can be woody, like trees or shrubs, or herbaceous, dying down to the ground each winter and regrowing each spring.

One such perennial is Turkish rocket. It is a herbaceous perennial equivalent of sprouting broccoli: it is in the same family and, like broccoli, the immature flower stems are eaten. The difference comes after the harvest has been eaten and the plants run to flower. At this point the broccoli is dug up and put on the compost heap. New plants then have to be raised from seed again, and tiny seedlings weeded, watered and defended from pests. The soil spends a period empty of roots, and any nutrients it holds are vulnerable to being washed out by the rain.

By contrast, once Turkish rocket stops being harvested the first thing it does is flower extravagantly. The flowers are a magnet for all sorts of pollinators, but particularly hoverflies. The larvae of some hoverflies eat potential garden pests such as greenfly, so the garden benefits wildlife and, in return, the wildlife benefits the garden.

Since the plant remains in the ground, its leaves continue to photosynthesise, pulling carbon dioxide from the air and water from the ground to make organic matter. Its roots also remain, ramifying through the soil and into its lower levels to make sure that not a scrap of fertility is lost. I once had reason to dig a soil pit next to my Turkish rocket and I was still finding roots at over 1.2 metres (3.9ft) down. This permanent root system, alongside an undisturbed ecosystem of soil bacteria, fungi and minibeasts, is the reason why forests hold on to nutrients so tenaciously and use them so efficiently. In my forestry classes I was taught that newly-planted forests will benefit from fertiliser, but once the trees have grown so that their crowns meet and their roots fill the soil there is no value to adding any more since they will use what they have so effectively. Tropical rainforests often grow on extremely nutrient-poor soil, weathered by millions of years of heavy rainfall. All the fertility is held within the living ecosystem, which is why their removal is such a disaster.

Towards the end of the season I generally cut down and compost above-ground parts of Turkish rocket, so it acts as a green manure, feeding other parts of the garden. Because the plant fills both soil and air for the whole growing season, weeding also comes largely for free: there is simply no space for weeds to take advantage of.

There are annuals in the forest garden, but perennials form its backbone, each of them with the advantages above (there is nothing special about Turkish rocket in particular). The annuals tend to be ones that self-seed, so they also go through their whole lifecycle.

Despite the benefits of perennials individually, the full advantages of an agro-ecosystem only emerge when you consider the system as a whole, with its qualities of diversity, complexity and connectedness.

The diversity of the system, with many different crops spread through time and space, means that it is difficult for a crop pest to find the plant it specialises in, amidst the jumble of colours, shapes and smells. If a pest finds a monoculture, it can quickly spread to the whole crop. Even if a pest finds a plant to its liking in the forest garden, the plant next to it is less likely to be.

Nature is not just aimless diversity, however: it also has complexity, or structure. Each plant has a preferred niche and some flourish in the shade of others. The perennial nature of the garden means that the plants can be fitted together like a four-dimensional jigsaw puzzle, putting more crops into the space and the growing year than would otherwise be possible. Perhaps this is why visitors regularly comment that my allotment seems larger than a normal one.

Finally, in a forest the parts of the puzzle are not separate but connected. In a natural forest the roots of the trees are joined to a network of fungi. While we may identify fungi with their fruiting bodies, the familiar mushrooms and toadstools, the real 'body' of a fungus is a mass of tiny underground filaments called hyphae, of which there can be more than a mile in a single teaspoonful of soil. Fungi that link up with plants are referred to as mycorrhizae, meaning 'fungus-roots'.

With their large leaf area in the sun, trees produce lots of sugar but struggle for nitrogen and other nutrients. With their much greater penetration of the soil the fungi have these in abundance but need sugars, so they make perfect trading partners. Each tree usually has many fungal partners and each fungus in turn connects to many trees, so the result is a 'wood wide web' capable of trading scarce nutrients from one corner of the forest to the other. I'm not aware of any research into whether these partnerships exist in food forests, but the relatively undisturbed conditions in the soil at least means that the conditions are there for them to form. Plants that are uprooted and soil that is turned over every year have no such opportunity.

The perennial nature of a home garden means that it differs from an annual garden in several important ways. One is the nature of the work it requires. Forest gardens are sometimes optimistically described as 'no-work' systems. This unfortunately isn't true, but it is true that the nature of the work is different: the focus moves from sowing, weeding and generally caring for tiny plants, to harvesting.

I find this more enjoyable, and the work schedule is usually more forgiving. I had a vivid illustration of this when I damaged my back in 2015. I was in hospital for over a week and couldn't get out to the garden for months. By the time I could do any work again my annual beds (I do have some!) were a complete write-off, the seeds that I had sown previously completely buried under weeds. The forest garden, on the other hand, had barely noticed that I had been away: the main difference was that it had grown a little larger. I had missed a lot of harvests of course, but this fertility had been reabsorbed by the system rather than being lost and wasted.

Harvesting from a forest garden feels rather more like foraging than picking from a traditional veg patch: in fact one way to think of forest gardening is that you are creating a compressed foraging resource on your doorstep. I rarely go down to the garden with a specific harvest in mind; rather I go to see what is coming up today and I usually come back with at least half a dozen ingredients, often many more. This in turn means that cooking from the home garden is different, to the extent that there is a whole chapter on the subject later in the book.

When harvesting annual veg, you are often picking the whole plant. With perennial plants you are only ever picking part of the plant, generally either a fruit or seed or a fresh, growing part such as a leaf or flower shoot or a stem tip. Being stuck in the one place for years, most perennial plants invest in defences such as bitter or poisonous chemicals, thorns, hairs, indigestible fibres or thick bark to deter animal pests. From the point of view of the plant, humans are just another animal pest, so these defences make the plants unpalatable to us too. The growing parts are the exception. Technically these parts are called the *meristems* and they are worth recognising.

On a linden tree (*Tilia*), the above-ground meristems are the shoot tips including the young leaves that they bear, and the entire surface of the tree under the bark. No other part of the tree is growing.

Both are edible. The shoots make an excellent salad. The tissue under the bark, which adds girth to the tree by laying down wood on the inside and bark on the outside and also serves to transport water and sugars around the tree, is a little-known treat. In asparagus, the meristem is the entire asparagus 'spear'. In broccoli it is the whole immature flower.

Meristems are more edible than mature tissues for several reasons. The plant can't lay down indigestible cellulose fibres until the part has finished growing, because cellulose fibres are the plant's scaffolding and, like scaffolding bars, they can't stretch and would be snapped by the expanding tissue. Protective structures like thorns and spines are also missing their stiffening and so are soft and useless. Proteins and carbohydrates are concentrated in the meristem because they are needed by the plant to build the growing tissues, while defensive chemicals are generally added later.

Meristems can be recognised because they are softer and bendier than mature tissues, and often lighter in colour. Meristematic leaves even stretch if you pull them, and are often carefully folded up. The rule of thumb for working out how much of a growing shoot will be edible is often to see how much you can easily snap off. This is the tender and therefore palatable part.

The focus on growing parts leads to three important forest gardening principles: 'Always one meal', 'The more you pick the more you get' and the 'Panda Principle'.

Why 'always one meal'? Our expectations of vegetable gardens are conditioned by annual veg plots where, during the harvest season, we see a lot of food available to harvest at once: rows of beans, lettuce, carrots, potatoes, beans – all of which could be picked there and then. Looking round a forest garden with these eyes, there might not seem to be so much palatable food available at any point: just lots of inedible perennial growth. This is true, but there is always one meal available, in the form of those tender growing parts. If they are picked, the plants will quickly grow more and there will be another meal tomorrow. If they are unpicked, they harden off and become part of the structure of the growing plant – which continues to grow more tender growing shoots. As a result, although it might sometimes look like there is only enough yield for one meal, there is *always* enough for one meal, which in the end is all you need. This is true not just in the harvest season of the annual garden, but throughout most of the year.

The second rule springing from the use of tender shoots is 'The more you pick, the more you get', which is a very pleasant thought. When we pick the growing tips of a plant, it usually responds by producing more. The very act of harvesting also serves to prolong the growth phase of a plant and delay the hardening off and reproduction phase. With plants such as lime or saltbush, which produce leaves for salads, it is important to pick the whole growing shoot tip rather than just individual leaves, as this is what prompts the growth of more.

Finally, what on earth is the Panda Principle? This refers to a mystery, recently solved, about the digestive systems of panda bears. Although pandas are highly evolved for feeding on bamboo, even possessing a 'thumb' evolved from a wrist bone for holding the shoots in place, their digestive systems look like that of any other, carnivorous, bear. Herbivores like cows and sheep usually have complicated adaptations like multiple stomachs to deal with their nutrient-poor food. This apparent contradiction was recently solved by a team at the Chinese Academy of Sciences that looked carefully at the nutrient profile of a panda's diet.[1] To their surprise, they found that a panda gets about the same amount of protein from its diet as a polar bear does. By specialising in bamboo shoots, the meristem of the otherwise very poorly digestible bamboo, panda bears manage the equivalent of a carnivorous diet. Our aim in the forest garden is to be a panda, not a sheep – but with a more varied and exciting diet than the one that pandas have yoked themselves to.

The forest garden year

To understand forest gardening, it also helps to understand how the forest garden year differs from that of the annual veg patch.

Over winter, the forest garden yields mostly roots and a few hardy plants that take advantage of any warmer periods to put on a little growth.

Spring is a very productive period as all the perennials start back into growth. Since they have stored nutrients in some form over winter this growth can be very fast. The result is a wide range of leaves and leaf and flower shoots, from the very mild (for salads and spinaches) to the strongly flavoured (as herbs and flavourings). An important group at this time is the 'spring ephemerals', plants that carry out all their growth and reproduction in the ground layer before they are shaded out by the trees leafing up.

By summer, most of this fresh growth has either died back or hardened off as plants turn their attention to flowering and fruiting (although some continue to produce fresh growth throughout the year, either continuously or in spurts). For fresh growth, and for some early roots, it is now the turn of the annuals, both traditional vegetables grown in the traditional way and those which self-seed around the forest garden. Summer is also the time for edible flowers and for a succession of fruit, from trees, bushes and canes to ground layer plants.

Autumn continues the harvest of fruit and seeds. Some annual crops have finished but successional sowing can ensure a continued supply. It is also the time when many perennials have a new flush of vegetative growth, taking advantage of any good weather late in the year once seeding is over. This means that autumn is the most diverse time of year, in terms of crops.

Season		Crops	Examples
Winter	Nov to Feb	Winter-growing plants (mostly biennials and winter annuals)	Annual leek, herb celery, wintercresses, kale, spiderwort
		Roots, tubers and other underground structures (from the ground)	Greenbriar, parsnip, horseradish, lilies, salsify, scorzonera, sweet cicely, burdock, caraway, skirret, potato, hopniss, sunroot, camas, ulluko, mashua
		Stores	
Spring	Feb to Jun	Roots etc.	Fawn lilies, fairy spud, turnip-rooted chervil
		Leaves	Wild garlic, snowbell, ramps, mountain onions, chives, other alliums, miners' lettuce and pink purslane, waterleaf, lesser celandine, sorrels, wasabi, wood violets, kale, red valerian, dittander, patience dock, sea beet, herbs
		Leaf and flower shoots or stems of trees, herbaceous perennials and biennials	Lime, harigiri, koshiabura, toon, aralias, greenbriars, bellflowers, fern fiddleheads, mayflowers, Solomon's seal, twistedstalks, nettle, honewort and mitsuba, nalca, bamboo shoots, poke, lovage, hablitzia, ground elder, sweet cicely, alexanders, burdock, butterburs, angelicas, bride's feathers, asters, kale, Turkish rocket, caraway, seakale and colewort, hogweeds, rhubarbs, spiderwort, fennel, daylilies, radish tops and seed pods, patience dock, skirret, sea beet, perennial leeks, mallows
		Seeds	Elm
		Pollen	Pine
Summer	Jun to Sep	Summer fruits	Bunchberry, strawberry, amelanchiers, summer raspberries, hybrid berries, plums, cherries, blue bean, blueberries, huckleberries, currants and gooseberries
		Flowers	Elderflower, bellflowers, marigold, nasturtium, mallows, salsify, artichoke, chive and other alliums, king's spear, daylilies
		Roots	Sweet cicely, burdock, caraway, salsify
		Growing tips, leaves and leaf stems	Hops, aralias, saltbush, sorrels, kale, fennel, rhubarb
		Self-seeding annuals	Mustard, sow thistles, rocket
		Mushrooms	Shiitake
		Traditional annual veg	
Autumn	Sep to Nov	Autumn fruits	Apple, pear, elderberry, bramble, autumn raspberries, salal
		Nuts and seeds	Hazels, hickories, walnut relatives, pine kernels, poppy
		Roots etc.	Greenbriar, parsnip, horseradish, lilies, salsify, scorzonera, sweet cicely, burdock, caraway, skirret, hopniss, potato, oca, yacon, sunroot, camas, ulluko, mashua
		Regrowth of herbaceous perennials	Bellflowers, honewort and mitsuba, lovage, ground elder, sweet cicely, alexanders, kale, seakale and colewort, hogweeds, spiderwort, fennel, patience dock, sea beet
		Mushrooms	Oyster mushrooms
		Traditional annual crops	

One thing you will see from this is that perennial and annual crops are complementary, as they fill in the gaps in each other's productive seasons. The time from February to May, when the last season's stores are running low but the new season's crops have not yet matured, is traditionally known as the Hungry Gap in annual growing. I find the Gaelic version of the phrase particularly evocative: *an sgrioban liath an earraich* – the grey scratchings of spring. In the forest garden, I call this period the Wildly Productive Gap, as it is when perennial plants are mobilising their winter stores into abundant new growth. There is, however, a Perennial Hungry Gap too, in high summer when most fresh growth has stopped and plants are concentrating on flowering and seeding, but the crops from those fruits and seeds aren't yet ripe. Fortunately, the annual plants slot into this gap perfectly. As a result I like to grow both annual and perennial crops. I'll describe how to integrate annual crops into a forest garden in Chapter 2.

Understanding your climate

My garden is located in Aberdeen, a coastal city in the northeast of Scotland. Aberdeen lies at a latitude of 57 degrees north, the same latitude as Gothenburg in Sweden, Nizhniy Novgorod and Yekaterinburg in Russia, Admiralty Island in Alaska and St John's in Newfoundland and Labrador. Like most of northwest Europe, it is warmer than its northerly position would suggest thanks to the Gulf Stream, an ocean current which brings heat up from the Gulf of Mexico.

In ecological terms, Aberdeen lies within the great temperate forest biome, which stretches across Europe north of the Mediterranean region, extends an arm east into Russia, reappears in Asia in northern China, Korea and most of Japan, and covers much of the eastern United States. In the southern hemisphere, temperate forest is found mostly in New Zealand and eastern coastal Australia. Outside these areas, temperate forest may also be found scattered in mountainous areas such as the Caucasus, the Himalaya and the Andes.

However, my city only just squeezes into the temperate forest zone. It is very much at the northern or cooler end. Go inland and upwards a short distance and you will cross into Caledonian pine forest, part of that other great northern forest biome, the boreal forest, which covers much of Canada, Scandinavia and northern Russia. Where temperate forest is usually mostly made up of broadleaves such as oak, beech, birch and maples, the boreal forest is dominated by conifers such as pine, spruce and larch.

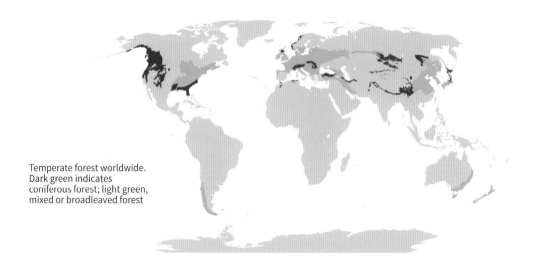

Temperate forest worldwide. Dark green indicates coniferous forest; light green, mixed or broadleaved forest

Average Annual Extreme Minimum Temperature 1976-2005		
Temp (F)	Zone	Temp (C)
-60 to -50	1	-51.1 to -45.6
-50 to -40	2	-45.6 to -40
-40 to -30	3	-40 to -34.4
-30 to -20	4	-34.4 to -28.9
-20 to -10	5	-28.9 to -23.3
-10 to 0	6	-23.3 to -17.8
0 to 10	7	-17.8 to -12.2
10 to 20	8	-12.2 to -6.7
20 to 30	9	-6.7 to -1.1
30 to 40	10	-1.1 to 4.4
40 to 50	11	4.4 to 10
50 to 60	12	10 to 15.6
60 to 70	13	15.6 to 21.1

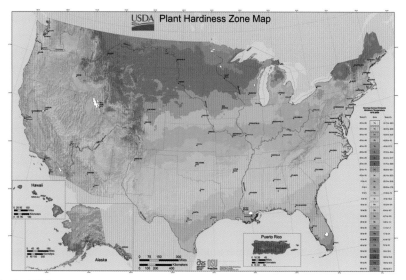

Plant Hardiness Zone map of the USA using USDA temperature ranges

When selecting plants to try in your garden, it is worth being aware of where it sits within the global matrix of climate and ecosystem type. This is because plants that grow in similar conditions elsewhere in the world are likely to do well in your garden. Equally, the similarity of your climate to mine will determine how useful the plant list in the last section of this book is for you. There are three main variables: winter temperature, temperature range and rainfall.

Winter temperature helps to determine what will survive in your garden without being killed by the cold. It is affected by latitude, altitude, how close you are to the sea or other large bodies of water, and local conditions. US Department of Agriculture (USDA) hardiness zones are based on winter temperatures and these have been adopted around much of the world. Almost all of the UK and Ireland fall into zones seven to nine. You can also measure minimum temperatures directly in your own garden with a min-max thermometer, in which two small plugs of metal are pushed round by the expanding and contracting mercury or alcohol and are left stranded at the minimum and maximum temperatures for the day.

The second big factor is the **temperature range** or summer temperature, which affects not so much what will survive as what will thrive in your garden. Areas close to the ocean, sea or giant freshwater lakes, like the Great Lakes of North America or Lake Baikal in Russia, have their temperatures evened out by the massive heat storage capacity of large bodies of water. Winter temperatures are milder and summer temperatures are cooler. This is known as a maritime climate. By contrast, areas far from the coast experience what is called a continental climate, with violent swings in temperature between winter and summer.

The mild winters of a maritime climate benefit many species of plants but bring their own set of problems. The cool, often cloudy, summers mean that plants can struggle to ripen fruit. There is also a parallel process of 'ripening' woody growth to make it frost-hardy by winter. Maritime climates are more unpredictable. In Aberdeen the first frost can come in September or December. Winter can bring three months of snow or none at all. The risk of plants getting caught by an early or late frost is much higher in a maritime climate. Also, ironically, plants in a continental climate often pass

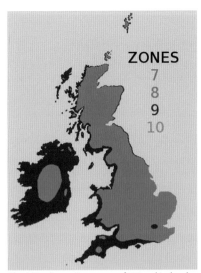

Plant Hardiness Zone map of UK and Ireland

the winter cosy under a thick blanket of snow. As warm-blooded animals we don't think of snow as warm, but it is a great insulator, shielding buried plants from the killing temperatures above it and keeping them to a few degrees below zero centigrade. In a maritime climate, snow comes and goes, whisking away the blanket when the plants need it most.

This maritime-continental difference means that hardiness zones are not an infallible guide to how a plant will do in your garden, particularly in maritime areas. I am often disappointed by plants with 'Siberian' in their name. Our stereotypical picture of Siberia is of a freezing hell, but plants that can survive the deadly winters of the continental climate reap a rich reward of hot summers and long, bright summer days. Plants that are at home there struggle to fruit properly in cool, dim maritime summers.

Plants from continental or maritime zones may also face unexpected problems when transplanted to the opposite area if they are waiting for particular cues such as day length or temperature to begin part of their growth cycle. One plant I have difficulty growing is mountain mint (*Pycnanthemum*).

It has no problem with the winters but doesn't start growing again until high summer, by which time it is generally too late. I imagine that this is because it is waiting for a set temperature to start growing, temperatures that come in early summer in a continental climate.

The final variable is **rainfall**. For obvious reasons this is higher closer to the sea, although some coastal areas affected by a cool ocean current, such as the Humboldt current which brings cold Antarctic water up the west coast of South America, can be unexpectedly dry. As you head for the continental interior the rainfall levels reduce. That this decline happens over a longer distance than the temperature effect is partly due to the forest itself, which sucks vast amounts of water out of the ground and returns it to the sky, where it is carried further inland to fall as rain again. Mountains can interrupt this process somewhat, forcing clouds to rise and drop their rain. Aberdeen is drier than most of Scotland because it is in the 'rain shadow' of the Cairngorm mountains.

Rainfall is the reason why, as we head east across the temperate belt of Eurasia, the vegetation turns from forest to steppe and finally to desert, reversing the sequence only as we approach the east coast of Asia. In North America the steppe is known as prairie, and researchers are working on a 'prairie agriculture' equivalent to forest gardening, using the perennial grasses and herbs of the region to create an edible prairie ecosystem. Fascinating as this is, it is outside the scope of this book and prairie plants will only be included where they happen to flourish in the forest garden, which, by and large, they don't.

Plants that do well in Aberdeen tend to come from the cooler end of the temperate zone or from the boreal forest. One of the closest analogues is the montane forest of North America's Pacific North West. Many of Scotland's commercial timber trees come from this zone, and fortunately there is a particularly long and rich history of plant use in this region to guide us.

Despite all this, however, plants can still surprise us, and some plants of the hot, dry Mediterranean zone, like artichokes or king's spear, can flourish here, far from home. In the end the only way to be sure whether a plant will grow in your garden is to try it.

The climatic location of your garden has consequences not just for the plants that you grow but for the forest structure that you will try to emulate. Robert Hart identified seven layers of the forest garden: dominant trees (the tallest ones), sub-dominant trees growing beneath them, shrubs, herbaceous plants and ground-hugging ground-covers, plus the root and vine 'layers'. You might see a forest with all these layers growing at the warm end of the temperate zone, but not at the cool end. In practice there are few shrubs that will grow productively in full shade and the herb layer and the ground-cover layer are fuzzy categories that I find easiest to treat as one (which I call the ground layer). Two layers are usually enough, plus the root zone. Vines or climbers are usually best grown up a frame in sun, rather than in the shade of a tree.

The small-scale forest garden

You don't need a large garden for forest gardening. Mine is in an allotment,* roughly 200 square metres (2,200 square feet) in size. Smaller spaces do impose some restrictions compared to larger ones, and might change your forest gardening style somewhat. Firstly, with less space, you won't want to waste any, and it is harder to fit in a diversity of crops. In a larger garden, labour may well be the limiting factor: in a smaller one, space is more likely to be. My main response to this is to focus harder on the ground layer, where individual crop plants are smaller, with fewer trees and large shrubs.

* For anyone not familiar with allotments, they are a small piece of ground, usually rented from the local government, for growing vegetables.

My allotment, 200m² of diverse food forest

Fortunately, this style is well suited to the northern climate and plant range as well. My ground layer is filled almost entirely with edible plants, doing without the plants for purely ecological effect that some forest gardeners use.

You also generally need to think more about your neighbours with a small garden. There might be restrictions on the height of plants that you can grow, either due to allotment rules or simply to avoid being a bad neighbour by shading adjacent plots too much. Similarly, any plants with strongly spreading tendencies are best kept away from the edge of the garden. I'll talk more about being a good garden neighbour in Chapter 3.

In turn, the surrounding spaces will probably impinge on your garden in some way. There is an elm tree on one border of my plot, which casts shade, drops twigs and leaves and has ambitions to turn my forest garden into an elm forest. You may need to adapt your forest garden design to surrounding trees, buildings and other features.

Forest garden products

This book concentrates on forest gardening for food, but there are many other possible products of a forest garden, including timber, poles and canes, basketry materials, medicinal plants, dye plants, soap plants, foliage and flowers. Each of these could be the subject of a book on its own and would make this one far too large, so I haven't covered either the species or how to use them. This is not to say that plants for these other uses shouldn't be grown in a home garden; any plant that you find useful should be included. The issues involved in growing the species and integrating them into the overall design are not significantly different, whatever the end use.

Another area that is not covered is the use of animals in the forest garden, but this too is possible, with a forest garden making an excellent site for a bee hive or poultry. Larger grazing animals pose more obvious conflicts with the plant crops, but tethered or penned animals are common in tropical home gardens.

How a forest garden grows a community

A home garden is not just a food-growing space, it is also a social space, for people as well as plants. In Indonesia, complex social rules attach to the use of pekarangans. They are typically open and other people can pass through – there is no concept of trespassing. Friends and family can enter the garden and help themselves to water from a well, dead wood or even vegetables; they seek permission in advance but it is rarely denied. Permission to take medicinal plants or materials for religious ceremonies is never denied. Researchers Otto Soemarwoto and G.R. Conway report that in some areas there is a belief that for a medicine to be effective it must be stolen and hence it is taboo to ask for permission. If the owner of the home garden discovers someone in the act of 'stealing' medicinal plants, she or he will pretend not to see them! The part of the home garden in front of the house is kept free of leaf litter, providing a place for children to play and for adults to meet for sport or to exchange news and views.

Cherries in boxes

We don't have similar traditions for temperate home gardens, but I can easily see from my own how they could develop. I often share surpluses with neighbours or take them to the local community centre for distribution. If I will be away, I actively encourage neighbours to go and harvest any crops that I will miss. The traditions around medicinal plants make sense as curative plants must always be available but are used only occasionally, and when someone needs them they really need them. A diverse set of medicinal plants scattered around the whole community and available to all is a sort of botanical National Health Service from which everyone benefits.

The children in my neighbourhood enjoy coming into the forest garden, as much for the network of paths to play chase around as for the food they can harvest. I encourage them to pick fruit. Compared to a regular garden, a home garden takes less work to maintain but more work to harvest, and doles out its yields in small amounts over the whole year, so I can be more relaxed than traditional gardeners about theft or over-harvesting. There is little opportunity for anyone to swoop in and carry off the fruits of months of work overnight. Anyone who wants to take a harvest from my garden will have to put in some work themselves. In a similar way, a forest garden is relatively robust against vandalism.

These qualities also make forest gardening a good approach for community gardens. Shared public gardens often present a challenge in how to share out the food they produce equitably, sometimes with the result that no-one uses the harvests at all! In a forest garden, since there are no harvests without effort and since there is usually more to pick a short while later, this problem takes care of itself to a degree. The challenge instead is that many of the species grown are unfamiliar to a lot of people.

Framed by elder: the Scottish Wild Harvests Association meeting at Cairn O'Mohr winery

The permanence of a home garden is both a strength and a weakness. It fits badly with the demands of a modern capitalist economy for a 'flexible labour market', with workers in precarious jobs on short-term contracts, ready to up sticks at a moment's notice and move to a new job in a new city, atomised and alienated from family, community and land. You can't uproot a forest garden and take it with you to a new job. Community forest gardens are one solution to this problem, as the community is more fixed and permanent than any of its members.

In the old expression *kith and kin*, 'kin' is family, but 'kith' means something like 'familiar country, place that one knows', and also 'friends, acquaintances'. Humans need a kith. We need to be grounded and rooted in families, communities, cultures and landscapes; we do not thrive as tumbleweed. A home garden might be a drag on our free movement in the flexible labour market, but it can also be an anchor, reminding us of our true needs and that we should not reduce all our relations to money or build societies where many are forced to live precarious, mobile lives to better serve the needs of capital. A forest garden is a place to be human in.

Lily growing out from the
shade of other vegetation

2

Designing Your
Forest Garden

If you want to plant a home garden of your own, it helps to have some idea of the final design before you start. This will only be a rough sketch, as designs can and should evolve with time, but it will give you a guide to work from. The steps involved in drawing up a home garden design are:

1. Survey the site.

2. Plan the location of the shady areas, the sunny edge, special niches and annual beds.

3. Decide where to put any infrastructure like sheds and plan a network of paths and beds.

4. Plan where to put individual trees and shrubs.

5. Fill in the ground layer.

This chapter covers the site investigation and mapping that you will need to do, the different niches that you can create, where to fit annuals into the system, how to stack different plants in layers, how to stop spreading plants from taking over everything else, and the importance of a good path network. In the next chapter we'll turn to how to implement the design on the ground.

Investigation and mapping

The first stage in designing a forest garden is investigating and mapping the existing conditions on the site.

Understanding your soil

The first thing to consider is the soil. You might have anything from a rich, deeply-worked allotment soil to a stony field. If your soil imposes constraints on what you can grow then you have three options. The first is to carry out major soil modifications such as double digging. These need to be carried out at the start before you have perennial plants in the ground. The second is to modify the soil by slower, gentler methods which take place over years while the plants are present (and are at least partly carried out by them). The third is to adapt your design, planting only what will cope with your soil conditions. This way you can make a virtue of the constraints, for there is a plant adapted to each one of them. For maximum diversity of crops you don't want a uniformly 'good' soil, but a mix of conditions supporting a mix of plants. In reality you are likely to take a combination of approaches.

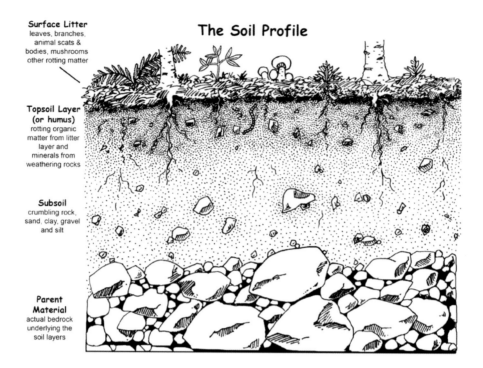

The Soil Profile

Surface Litter leaves, branches, animal scats & bodies, mushrooms other rotting matter

Topsoil Layer (or humus) rotting organic matter from litter layer and minerals from weathering rocks

Subsoil crumbling rock, sand, clay, gravel and silt

Parent Material actual bedrock underlying the soil layers

Soil profile

Soil is a three-dimensional structure, so beware of thinking of just the top layer as 'the soil'. In a new garden I like to dig at least one soil pit and have a good look at the soil profile: the different layers of the soil as you go down.

All soils are formed from an original 'parent material', which can be the underlying bedrock of the site or materials brought from elsewhere by wind or water, such as sand, gravel, silt or loess. Over time, plants, animals and the weather act on the parent material to turn it into living soil. Because these influences are strongest at the surface, we tend to think of soils from the top down.

On top of most soils there will usually be a thin layer of organic material that has not been broken down yet, such as leaves and twigs. Below this there is the topsoil, a layer that contains the break-down products of years' worth of organic matter, called humus. The presence of humus makes the topsoil darker and usually more crumbly than the lower layers.

Below the topsoil is the subsoil. The subsoil is like the topsoil in being visibly altered from the parent material by the action of roots and worms, but it contains little organic matter. The subsoil may contain lots of minerals, such as iron, washed out of the top layers, and this may give it a reddish colour. In undisturbed soils in areas of high rainfall and acid conditions (such as heaths and conifer forests in Scotland) there may be a whitish layer between the topsoil and the subsoil, where all minerals have been washed out into lower levels. This is called the 'eluviated' or rained-out layer. In soils that have been previously worked these layers may be mixed up and hard to distinguish. Below all this is unchanged parent material.

From a gardening point of view, the most important thing in a soil profile is whether there are any impermeable layers which will impede rooting and drainage. Where this is bedrock there is little that can be done, but other impermeable layers can be broken up, greatly improving the soil. Breaking up these layers means that roots can grow deeper into

A soil with an eluviated layer. The grey band between the two blacker layers has had nutrients and organic matter washed out of it by slightly acidic water draining from above.

the soil and will bring minerals up from the lower levels to which they tend to get washed.

Many northern soils have been affected by glaciers in the past. The immense pressure produced by these leaves a highly compacted layer that water and roots cannot pass through. The compacted layer is later covered by other materials so it appears at a lower level in the soil than the surface. They may have less compacted material underneath them, so breaking up the layer can increase the effective soil depth considerably. Compaction can also be caused by human activities.

Another kind of impermeable layer is called an iron pan. This is where iron washed out of upper layers precipitates out at a certain depth, concreting the soil sediments together. An iron pan is visible as a rust-red line in the soil. The line can even run through the middle of stones that sit at the level of the pan! In ex-agricultural land, another kind of pan is called a plough pan, formed by the plough repeatedly smearing clays at the bottom of its depth.

All these kinds of impermeable layers can lead to water sitting on top of them in what is called a perched water table. The wet conditions lead to a low-oxygen zone which becomes grey and may even smell anaerobic (like a blocked drain). This is called gleying. A gleyed layer can also be caused by a high water table due to other causes, such as being in a hollow.

Soil texture

Besides the soil profile, the most important characteristics of soil are texture, pH and structure. Soil texture refers to the mix of different sizes of soil particle in the topsoil. These particles may be sand, silt and clay. Soils that contain a lot of sand feel gritty. They are light and easy to work, but poor at holding onto water and nutrients, so they are not very fertile unless both of these are added continuously. Silty soils feel soft and silky. They are easy to work like sandy soils, but better at holding onto water. Their downside is a tendency to form a hard crust in dry conditions. Clay soils feel sticky. The ion-exchange sites on the surface of clay minerals allow them to bind lots of water and minerals, so clay soils are generally high in both, which makes them very fertile. The downside is that they can easily become waterlogged and are heavy and difficult to work. The ideal is generally considered to be a mix of all types, with 40% sand, 40% silt and 20% clay. This is called a loam.

The magic ingredient in all soil types is organic matter. Organic matter holds onto water and nutrients in a similar way to clay particles, so it makes sandy soils more fertile. At the other end of the spectrum it opens up the structure of clay soils, making them less sticky, easier to work and more freely draining.

You can test your soil texture easily, so do plenty of tests across your site. In all cases you want to test the mineral portion, not the organic layer on top, so dig down around 10cm (4in) to take a sample.

The simplest method of all is the touch test. Spread your soil sample out (squash it if it is really clayey!), remove all small stones and crush any little lumps of soil that appear to be holding together. Moisten the soil to help it hold together (don't make mud) and make a ball about 3cm (1in) in diameter. If it won't make a ball, even if you try a few times and adjust the amount of moisture, or if the ball falls apart when you toss it up and catch it gently, you have a sandy soil. If it makes a firm ball you can do further tests.

Next roll the ball into a sausage shape, 6-7cm (2.3-2.7in) long. If it breaks up in the process, you have a loamy sand, which is a sandy soil with small

Soil test doughnut – clay soil

amounts of silt or clay. If not, keep going until the sausage is 15-16cm (5.9-6.2in) long. If it breaks up at this stage, it is a sandy loam: a sandy soil with larger amounts of silt and/or clay.

If it survives again then you don't have too much sand in your soil. Now try bending the sausage into a half-circle. If it breaks now, you have a silty soil or loam. Lucky you! If not, continue to a full circle. If it breaks up it is a heavy loam, with a large amount of clay, or a sandy clay, about half sand and half clay. If it is the latter, it will feel gritty when rubbed between thumb and finger. If it makes a doughnut shape without breaking up it is clay soil. Add organic matter or take up pottery.

There are many variations on this test, which you can find on the internet. Another method which is almost as easy and which gives more precise results is the jar method. To carry this out, take and prepare a soil sample as before and use it to fill a straight-sided jar to about one-third full. Then fill the rest of the jar with water, leaving a small space at the top, put the lid on and shake well. This test relies on the fact that heavier particles settle out of the shaken mixture faster than lighter ones, and you can measure the proportion of sand, silt and clay by recording the amount of material that has settled to the bottom of the jar after different times.

The first to settle out is sand. After one minute, mark on the jar the top of the settled material. This is your sand fraction. The next is silt. After two hours mark the top again. This is your silt fraction. Clay is so light that it takes considerably longer to settle out. Mark the level after two days. You might find that it takes even longer than this: adding a little dishwasher detergent or borax will speed things up or you can just wait. Once the clay layer has settled you can measure the height of the marks on your jar and work out the proportion of each particle in your soil from the proportion that each makes of the total height of the settled material.

With the proportions of sand, silt and clay you can figure out what type of soil you have using the

Soil texture triangle

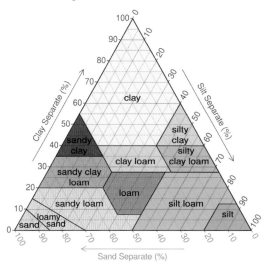

soil texture triangle. Follow the lines from each percentage; the soil texture is where the lines all meet.

Soil pH

You will also want to know the pH of your soil: how acid or alkaline it is. All plants have a preferred pH, so knowing that of your soil will give you a guide to what you can plant and what changes you might have to make to plant other things.

You can buy pH testing kits at garden centres, but it's cheaper just to buy a roll or pack of pH paper, which changes colour according to the pH of the solution it's dipped into. Look for 'universal' paper which spans the pH range from 1-14, not ones that just show acid or alkali. A cheap method means that you can measure your site in greater detail and over time.

To measure soil pH, dig a hole 30cm (12in) deep. Using a trowel, take a slice of soil from the side of the hole, so that you get a mix of soil from all depths (but avoid getting any surface organic matter). To make a composite sample that evens out random differences in small patches, take several samples: up to 15 in a large area that you think is likely to be uniform. Put the samples together in a clean bucket and mix well. Do a separate composite sample for each area that may have had a separate soil history (that has an obviously different soil texture or structure, or might have been treated differently, for instance by liming, which creates a more alkaline pH).

Put a little of the mixed soil into a small glass or plastic container and fill with distilled water to the same level as the soil. Distilled water has a neutral pH that won't affect the test result. It is sold for use in steam irons and is reasonably easy to get hold of. Tap water is usually not neutral and rainwater is always mildly acid. You can make your own

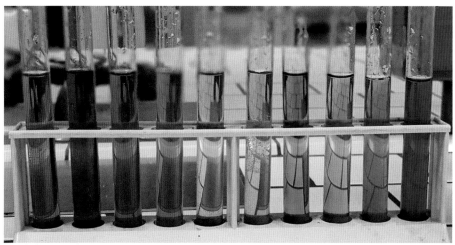

A pH indicator scale

distilled water at home if necessary. Swirl the sample around for a minute or so to mix well, allow it to settle for about 30 seconds and then pour through a filter paper (coffee filters will do nicely). Dip the pH paper into the water that comes through and compare the colour to the chart that comes with the paper. This will give you the soil pH.

Nutrients

Levels of various soil nutrients can also be measured, but nutrient levels are more of a function of your management of the soil and less of a property of the soil itself. The major nutrients are nitrogen, phosphorus and potassium. Nitrogen is important for plant growth and a component of chlorophyll, the green pigment that plants use to capture light. Phosphorus plays many roles in cell function, and potassium is particularly important for flowering and fruiting. If you do want to get these tested it is generally easiest to send a soil sample to a soil lab. Tests are not expensive and some labs even offer tests to local gardeners for free.

Drainage

The final characteristic of soils that you need to observe is drainage. Drainage depends partly on soil texture and profile, but also on where your site is in the landscape. All else being equal, a site in a hollow will be wetter than one on a slope. Drainage is harder to measure than other properties, but easy to observe: just notice how wet the soil is under different conditions. The plants growing on the site may also give a clue as wet areas tend to develop characteristic wet-ground vegetation. This can be misleading however if the soil is regularly cultivated: characteristic cover takes a while to develop.

Understanding light and warmth

Above ground, light and warmth from the sun are the most important things for your plants. The sun is at its highest when it is in the south,* and on average plants will get more light from this direction, so a plant on the south side of a shadowing wall or bush will get more light overall than one on the north side and land that slopes to the south will get more light than land that slopes to the north. This simple picture is complicated by the changing path of the sun throughout the year at high latitudes.

In winter, the sun will rise in the south east, reach a point low in the south by midday and set again in the south west. The further north you are, the further south the rising and setting sun shifts and the lower it rises. At the extreme, beyond the Arctic Circle, the sun doesn't even make it over the southern horizon in the middle of winter. If you think that your garden doesn't get much sun, bear in mind that Stephen Barstow, author of *Around the World in 80 Plants*, grows a perennial vegetable garden in Norway in exactly these conditions.

In the high latitudes, our reward comes in summer, when the sun rises in the north east, climbs high in the southern sky and sets in the north west. Plants can therefore get a significant amount of sun from the north during the growing season if the place they are in is open in that direction. Map the compass directions and the features that cast shade onto your growing space, but bear in mind all the different directions that light can come from at different times of day, in different seasons.

Sun direction is also significant for placing plants that are sensitive to frost. It is often not the absolute temperature that kills plants but the rapid thawing when the sun comes up. Sensitive plants should be placed where they are not open to the south east and exposed to the rising sun in winter.

Finally, walls and buildings may protect plants from frost by acting as a heat store, taking in warmth during the day and releasing it at night, raising the temperature a few crucial degrees for plants at their base. In the case of badly-insulated buildings, like the granite tenements of Aberdeen, they may even be an active heat source!

* In the northern hemisphere. Flip north and south for the southern hemisphere.

Planning your niches

Once you have got a feel for your site, you can think about the best places to locate different niches for your plants to grow. You can either create the different conditions from scratch or take advantage of existing conditions. For instance, there is no point labouring to create a wet bed from nothing if you have an area that is already low lying with poor drainage. The more niches we have in the garden, the greater the range of plants that we can grow in it.

The niches that you might want to consider creating include:

- Deep forest
- Forest edge
- Full sun
- Acid soil
- Alkaline soil
- Dry / free-draining areas
- Wetland beds and ponds
- Seashore bed
- Grass / amenity area
- Annual beds

The degrees of shade and sun will be created partly by the walls, fences and other structures that cast shadows in or into your garden, but also by the

Degrees of shade under trees

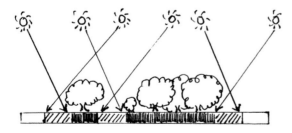

- deep shade
- forest edge
- full sun

plants themselves as the trees and shrubs cast shade onto the ground layer. **Deep shade** occurs where plants get little or no direct sunlight at any time of day. It is found under the canopies of trees or shrubs where they are planted close together so that their crowns meet, and around the base of larger-crowned trees even when they are standing on their own. Plants that flourish in deep shade are found in Chapter 9. Some plants, like wild garlic, last far better if they never get direct strong sunlight, so make sure you identify a few places with good deep shade for plants like these. On the other hand, there are fewer plants for deep shade than there are for more open niches, so don't squeeze in lots of trees and then discover that your whole ground layer is in deep shade.

Forest edge or open forest conditions occur where shrubs and ground layer plants are shaded for part of the day but receive direct sunlight for part of the day. They are to be found around trees where they are not planted so close together that their crowns join up, or where they have particularly open canopies. Since sunlight comes from an angle, ground layer plants under the crown of a tree can also receive sunlight for part of the day if conditions are open on the side where they are planted. The majority of temperate home garden plants are forest edge plants, so most of your trees should be planted sufficiently far apart to provide the right conditions for them. Spacing can be achieved either by pruning or by ensuring that each tree has more space around it than the natural spread of that species.

At the other end of the spectrum, some plants will only thrive if they get **full sun** all day. These are easiest to accommodate on the southern edge of the 'forest' area, away from any trees.

Similarly, there is a spectrum of **pH** requirements. Most plants prefer a pH in the range of 6.0 to 7.0, but a few do better in more acid or more alkaline conditions, and it is worth having a small dedicated bed for each. Acid loving plants are mostly in the heath family. An alkaline bed can be created using basic rocks like limestone or by adding lime or wood ash to soil. Lowering pH is harder. For fast results you can dig in iron sulphate; around 500 grams per square metre (1.6oz per foot squared) will reduce the pH by one. Elemental sulphur has a similar effect, but is slower acting as it needs soil organisms to convert it to sulphate. Alternatively, you can simply mulch the bed every year, especially if you use materials like pine needles. This will acidify the soil over time.

Plants differ wildly in the amount of **moisture or drought** that they need or can tolerate. A particular killer of perennials in cool maritime areas is winter waterlogging. You can bring plants through the winter that would otherwise perish if you give them a well-drained soil. I create a range of drainage conditions using 'half raised' or terraced beds. My beds are raised by up to a metre at the northern end but not at the southern. This has a number of beneficial effects at once. The whole bed is tilted toward the south, equivalent to moving it a few degrees of latitude towards the equator. A path runs along the base of each bed, so the shadow that is created falls on the unproductive path rather than the growing area. The raised area alongside the path provides an area that is particularly easy to weed and pick from, for small, fiddly plants like alpine strawberries. And it creates a spectrum of growing conditions, from very freely draining at the top to more moisture retentive at the bottom.

If you want to go further and provide conditions for true **wetland** or water's edge plants then in most gardens you will have to actively impede drainage to make a wetland bed or small pond. You can make a large pond but I won't cover this as most people don't have the opportunity and it really needs a book in itself. You can impede drainage over a small area using a solid barrier, a flexible one or a soil-based one. An old bath, sunk into the ground, makes a perfectly good solid barrier, or you can buy purpose-built moulded ponds. Concrete ponds may sound like a good idea but they have a tendency to crack in winter.

The crucial thing about a pond is that animals (both pond-dwelling ones that leave at some point, like frogs, and ones that fall in accidentally) need to be able to walk out. Moulded ponds are usually made with a suitable profile: shallow, preferably sloping, at one end or both. For something like a bath you need to add rocks and gravel to give a ramp up to and over the lip. This needs to be done well to avoid any risk of settling in the future. Besides making your pond animal-safe, this again increases the number of niches for plants, from deep water to margin.

You can also buy flexible pond liners. Here the shape must be formed from the underlying soil before the membrane is laid. This is usually done in several layers. The first layer, which can be geotextiles or sand, is there to cushion the waterproof layer and protect it from being punctured by rocks or roots. The waterproof liner goes over this, followed by another protective layer. If the inside protective layer is absorbent, it is important that it does not go up and over the waterproof layer. If it does it will wick water out of the pond and it will mysteriously go dry!

Finally, you can use the soil itself to create a more natural impermeable layer. If you have a high clay content in your soil, you can make your own lining by a process called puddling. This is how our original canal system was waterproofed, and that has lasted pretty well. The idea is to compact and smear the clay, driving out any air and making a resistant layer. This is achieved by donning the biggest pair of wellies that you have and treading the clay thoroughly while it is wet. As you might imagine, this is a very messy process. If you don't have a heavy clay soil you can buy 'puddling clay' to line a pond.

If you have an on-site or nearby source of clay, clay ponds are cheap and low impact and have the added benefit that small holes or cracks in the layer will self-seal as water flows in and the clay expands. On the other hand, they are vulnerable to roots growing through them and if the water level

Even a small pond can bring frogs into your garden

drops significantly they dry out and crack. They are only worth using if you can be sure that you can keep the pond constantly full.

The water level in other kinds of ponds may not be as critical as that in a clay-lined pond, but all ponds need topping up from time to time. Consider placing yours near your water butt so that this part is easy. In addition I have an overflow on my water butt that runs through a buried hose straight into the wetland, taking the water where I want it automatically.

One niche that might not immediately spring to mind for a forest garden is the **seashore**, but a lot of useful plants originate there. See the 'Coast' section of Chapter 11 for a list. Seashore plants are salt tolerant with adaptations to drought such as reduced, waxy or silvery leaves. I have one bed that I mulch with seaweed periodically to mimic the conditions on the storm line at the top of a beach.

Integrating annuals

I also recommend having some areas for annual crops, since there is no reason not to have both annuals and perennials. They complement each other in terms of the growing year and extend your range of crops. There are two approaches to growing annuals in a home garden. The first is to segregate them as far as possible from the perennials, in a dedicated annuals bed.

The main reason for this is to protect them from snails. Perennial crops both tolerate and harbour a higher population of snails than annuals (although see Chapter 4 for ways of managing your molluscs). This isn't a problem, unless the two are right up against each other, allowing the slimy marauders to hide out in the perennials and launch nightly raids on your tender annuals. The difference in the timetables of the two kinds of plants that is so beneficial in terms of harvesting then becomes a weakness, since the perennials are already in lush growth and offering hiding places when the annuals are at their most vulnerable.

My solution to this is to use the sun-loving perennials bed as a buffer between the more snail-friendly shaded beds and the annuals. The sunny bed is to the south of my 'forest' and the annuals are to the south of that, giving them both snail protection and full sun.

However, this approach is not necessary for all annual crops. Some flourish quite happily in amongst the perennials and can be grown in little clearings in the perennial beds. All they need is some bare soil at germination time. The seeds of many annuals are in fact naturally adapted to lying dormant in the soil for many years, just waiting for a gap in the perennial cover that they can spring into action and exploit. Annual crops that can be grown this way fall into two groups. They may be natural or near-natural species, little altered for cultivation, that have kept their adaptations to surviving in a tough world. An example of this is sow thistle. They may also be more familiar crops that have some resistance to snails either due to defensive chemicals and hot flavours – such as rocket – or large propagules that allow young plants to get established quickly, such as broad beans or potatoes. Plants that are suitable for growing in the perennial beds are included in the plant chapters; ones which need to go in the separate annuals bed are not.

I have been referring to annuals, but a lot of crop plants are really biennials, growing in their first year and flowering and dying in their second. If you want to maintain either annuals or biennials then obviously you will need to allow at least some plants to flower and seed. You can either collect the seeds for deliberate sowing or allow them to fall for self-seeding. After a while you will find that your 'weed' seed bank is largely edible plants.

You can leave spaces for seeding crops deliberately, but you will also find that ordinary garden management creates a degree of soil disturbance as you move plants or get rid of ones that you decide you don't want, or they simply die on you. Self-seeding annuals and biennials are a great way of making the fullest use of these accidental spaces. Gardeners usually try to 'rotate' annual crops, moving them from bed to bed to avoid the build-up of disease. With temporary spaces this happens automatically, but where you maintain and sow open spaces deliberately it is a good idea to move the crops around every year.

You can use space in the seeding beds most efficiently by starting some annual crops in pots, trays or a nursery bed. Many biennial crops, such as sea beet or celery, are most productive while they are starting to run to seed, producing flush after flush of nutritious flower heads. Once they truly run to seed, in late June or early July, I pull out all but the best, which I leave for seed collection. This leaves lots of small gaps, which are perfect (after digging in a little compost) for planting out crops like leeks and squashes which have been started off indoors. That way there is barely a break in the growing season.

Space for you in your forest garden

You might also like to consider a grassy amenity area within the garden, for sitting and enjoying the results of your labours. This is the one place in my

Potatoes growing in the corner of a perennial bed

allotment that I tolerate grass, as it makes a hard-wearing, self-regenerating surface – also known as a lawn. It is not primarily a food-growing area, although a few crops do manage to naturalise in the grass and the cuttings from it go to mulch growing areas. I try to minimise the length of the boundary between the grass and any beds, as it always takes work to stop the grass from invading them. I have my lawn on the southern edge of the garden, beyond the sunny beds and the annual beds, taking advantage of the open nature of the area to avoid shading the beds beyond. It is small enough that I 'cut' it literally by hand, pulling hand-fuls of grass somewhat like a browsing cow, to favour the wild flowers and avoid strimming the frogs that like to hang out there. Although small, it is a crucial part of the overall garden and a wonderful place to chop wood in winter or laze in the summer, listening to the birds and watching the clouds of pollinating insects at work.

Paths and access

The best crop plants in the world are not much use if you can't get to them when you need them. This is especially true in a forest garden where you crop a little bit regularly rather than digging up a large harvest all at once. You still want to be able to pick after a rainstorm on a summer's day when the plants are grown up on all sides, dripping wet and cold. This is where a really good path network comes in, and you need to plan for one from the start. We will consider materials, layout and how to make sure that even the paths play a role in the ecology and carbon capacity of an edible ecosystem.

Materials

Allotment paths are traditionally grassed, which must be about the worst surface you can think of. They are moderately hard-wearing in summer, but

Looking into the amenity area

Sitting area with essentials to hand

quickly go to mud with heavy use in winter, and they take constant work to maintain and to prevent from encroaching into your growing spaces. There are many, far-better options, all of which can play a role in your overall path network.

Slabs are hard work to lay and have a high embodied energy, but they do have the advantage that they are extremely hard-wearing and need very little maintenance once they are in place. They are best used in the areas of high foot traffic where these benefits will be most worthwhile.

Woodchip is the workhorse in my path network. A woodchip path is very hard-wearing so long as the layer is thick enough (this varies with soil type and wetness, but 5cm (2in) is generally enough – more than this is good if you can get it and will help to suppress weeds underneath it when you first lay it). It is not a no-maintenance surface: its virtue lies rather in the fact that maintenance is fast and very easy. Being soft and loose it is simple to

hoe, and a few times a year is enough to keep it weed-free once it is established. It does need to be topped up periodically, although less and less as time goes on.

Don't, whatever you do, be tempted by the idea of laying a membrane covered in woodchip. This may stop weeds from coming up from underneath, but these are not a problem after the first few months (or if you have done your preparation properly) anyway. The problem is plants rooting into the woodchip, which is low in nutrients but otherwise an excellent rooting medium for young plants. The membrane gives something for plants like grass to get their roots into that is very hard to get them out of again as it impedes the hoe. At best the membrane is unnecessary; at worst it is the weeds' ally.

If you buy woodchip in plastic sacks from garden centres it is prohibitively expensive, but there are far cheaper sources. Tree surgeons are usually

Two paths diverged in the forest garden … Woodchip paths are hard-wearing and easy to maintain.

happy to drop woodchip off for free if they are working in an area rather than trailing it a long distance home. This needs to be made as easy as possible for them: if you can designate an area where a tree surgeon can drop a load when they are in the area with no need to call anyone or get special access, then you will likely find loads appearing there on an unpredictable basis, as if the woodchip fairy has been. If you don't have such a space you might still be able to have a load delivered for a good price. Local authorities usually have piles (literally) of the stuff and if your allotment site or community project doesn't already have an arrangement for woodchip to be dropped off periodically, see if you can start one.

Stone chip has the advantage over woodchip that it doesn't degrade or need to be topped up, but it is harder to hoe and most stone chip paths end up being maintained chemically. I don't use it.

Earth paths are the simplest kind possible: simply hoe a line and there you have it. Their drawbacks are that they become compacted and harder to hoe over time, and muddy when wet.

Boards laid over beds make another simple kind of path, and a very good one. They help to protect the soil from compaction, they suppress weeds beneath them and they provide a refuge for all kinds of beasties, including beneficial ones like beetles. It can also provide a refuge for less helpful invertebrates, like flatworms or slugs, but this is beneficial in a way too: simply turn over the board and you know where to find them. Avoid chipboard or laminated or painted boards, all of which will contaminate the soil when they break down. Sawmill offcuts, or 'backs', work well.

Weed-mulched paths are something of an art, but with practice they provide a path, dispose of waste and feed the soil all at once. If they are not deep enough they become weed reservoirs and if they have too much 'green' organic waste (see Chapter 4 on composting) they can become rather sticky, but in between they are excellent. They make use of the fact that a lot of the organic waste in a forest garden is long, thin and somewhat woody, like raspberry canes or bolted biennials. These are awkward to compost conventionally, but easy to lay on a path, where their fibrous nature makes the path robust and the trampling helps to break the material down.

A path doesn't even have to be continuous. In places I sink in **stepping stones** instead, giving somewhere to put my feet without stepping on plants or soil. These need to be spaced well apart to avoid creating weed havens.

Layout

The layout of your path network needs to give access to all parts of the garden and to let you get from one part of the garden to another without taking complicated detours. Wide paths also make effective weed barriers, helping you to control plants that spread by runners. While you can make the layout as intricate as you like, a simple grid system is the most efficient solution in most circumstances. You need a network of 'main roads' for getting around easily and to break the space up into distinct beds. In my garden these are mostly made from woodchip, with a short slabbed length near the shed. A path runs

A

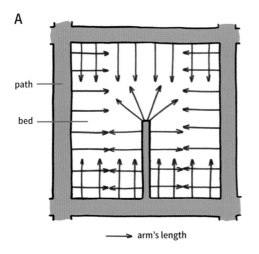

path

bed

⟶ arm's length

B

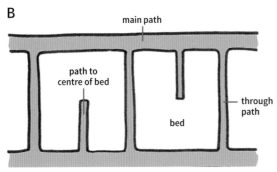

main path

path to
centre of bed

through
path

bed

C

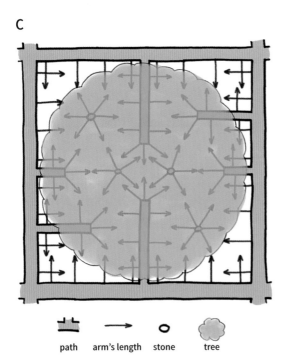

path   arm's length   stone   tree

A An access path to the middle of a bed, showing an arm's length reach into all parts of the bed

B Path options with terracing

C A large bed with a tree in the middle

around the perimeter, then the main paths largely run east-west, to fit in with the terracing.

Smaller paths then give you access into the beds. I use a combination of boards, weed-mulched paths and stepping stones. For maximum space efficiency, these paths only need to go to the middle of a bed. For greater convenience, it is often better to run them right through. The size of your beds is determined by the length of your arms, since the purpose of the bed layout is to allow you to be able to reach all parts of the bed without stepping on the plants. Larger beds can be created by running an access path halfway to the centre from four directions.

Paths have a tendency to disappear as vegetation grows up on either side of them. To keep them open in the middle of summer, make the main paths wider than you think you need (mine are about a metre) and taper your planting back from the paths, with the smallest plants closest to them and the largest in the middle of the bed.

Path edging often creates a refuge for weeds. There is no real need to edge a path or bed at all, but if you do then make sure that the edge you create can be hoed.

Deep wood paths

Paths can also be used as a solution to the amount of woody waste generated by a forest garden. A 'deep wood' path uses this waste, stores carbon, holds moisture and creates a lovely surface to walk on. I have converted most of my main paths to this type.

To create a deep wood path, dig a trench the width of your path, as deep as you like. The soil can be used for building raised beds. Fill it with any kind of woody waste and top off with woodchip. Build it slightly higher than you need as it will settle a little. Over time it will settle further and you will occasionally need to add another layer of woody

Digging a trench for a deep wood path

Filling the trench with waste woody material

waste and woodchip.

The wood under the path slowly breaks down to create a sort of 'wood mould', rather like the leaf mould that is made for potting. This holds water well and the path will help to hold moisture in the system, sucking it up like a sponge when it is wet and releasing it slowly in dry periods. When I have dug out an old path I have noticed that nearby trees root into them enthusiastically. Eventually you can dig out the path, use the wood mould and start again.

Planning your planting

In planning your beds, you can fit more than one crop into the same space, a principle often known as stacking. You can stack plants in three main ways: in layers, vertically; in time, by different seasonal patterns; and in the herb layer, by different space requirements.

Ecologists distinguish three different plant **layers** in temperate woodland: the canopy, the understorey and the herb layer. The canopy is made up of the tall trees, of which the 'dominants' are those which have managed to get ahead of the competition and bask in full sun over their whole crown, while the 'co-dominants' and 'sub-dominants' play catch-up, slowly being squeezed out, but ever hopeful that one of the dominants will fall and give them elbow room.

The understorey lies below the canopy but well clear of the ground. It has two components. Young trees of canopy species grow slowly in the shade, waiting for a break in the canopy to give them their big chance. Scientists have recently discovered that parent trees can feed their offspring through the fungal network to sustain them through this difficult period. The other members are shrubs and small trees that are adapted to a long-term life in the shade, like hazel, holly and hawthorn. Although these trees can grow in the shade, they do most of their fruiting and reproduction when a hole in the canopy opens above them and they experience full sun for a period.

Below this is the herb or ground-cover layer. Usually this consists mostly of herbaceous perennials which die down and regrow each year, but in ecosystems like the Caledonian pinewoods it can include low woody species like heather, cowberry and blaeberry. Below even the herb layer we find the forest floor, a layer of rotting leaves and wood, dominated by fungi and mosses.

A home garden is not identical to this, but follows similar principles. It has a tree layer or canopy, which is generally much lower than that of a wild forest. Unless you have a very large space, most of us do not have room for large canopy trees like oak, pine or sycamore. Food trees in the cool temperate zone tend to be small, like hazel, apple or devil's walking stick. Even where they have the capacity to be large trees, like linden, toon or hari-giri, we will often want to keep them smaller for space reasons and to keep the harvest in reach. The principle of 'the more you pick the more you get' only applies if you can pick the majority of the shoots on a plant, so trees where the shoots are harvested are more productive as well as easier to crop if they are kept within reach.

This means that trees that would be understorey species in wild forest, like hazel or elder, are better thought of as part of the home garden canopy. In turn, this restricts the home garden understorey to shrubs and cane fruit like salal or raspberries. I also include some climbers and very large herbaceous perennials in this middle layer for practical reasons.

Finally, the ground cover or herb layer of a home garden should be thicker than that of a wild forest. At my latitude the ground layer in a closed canopy forest is often quite sparse. In a home garden this would be a huge wasted opportunity as there are more potential crop plants in this layer than in any other. There are some layers of the natural forest that we simply don't need to duplicate. There is no benefit in having struggling sub-dominants in the canopy or a population of just-surviving sapling trees waiting in the understorey. Omitting these lets through more light, allowing a lush and productive ground layer.

Stacking increases yields, but there are diminishing returns to trying to fit ever more layers into the same space. I generally find that in any one spot two layers are enough: trees and ground layer plants; trees and shade-tolerant shrubs; or sun-loving shrubs and the ground layer.

Plants in the ground layer can also be stacked together in **time**. Some of them, called spring ephemerals, have evolved to pack their whole annual cycle into the first months of the spring,

making use of the sunlight on the forest floor before deciduous trees leaf up. Others stick around through the summer, putting up with lower light levels for warmer temperatures. These kinds can be stacked into the same space with little loss to either. An example is fawn lilies with wild strawberries. Other spring ephemerals are wild garlic, ramps, snowbell, victory onion, lesser celandine, pink purslane, miners' lettuce, pignut and waterleaf.

Ground layer plants can also sometimes be fitted together when their use of space both above and below ground is very different. Some plants, such as fawn lilies, mayflowers, Solomon's seals, leeks, burdocks and scootberries, are very 'up and down', with deep roots and upright leaves or stems. Others are both shallow-rooted and low-

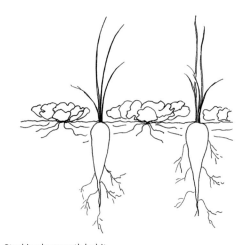

Stacking by growth habit

growing, such as violets, wild strawberries, Serbian bellflower, bunchberry, ground ivy, wintercresses and clover. These can be grown together without interfering with each other much. Some forest gardeners even treat the two as separate layers, but most plants fit in the middle of the spectrum; it is not always possible or desirable to have two in the same space.

Nettable beds

The need to net some of your crops (see Chapter 4 on pests) will also influence your layout. It is hard to net individual bushes scattered around the garden and intergrown with ground layer crops. It is easier to group plants that need netting together where you can put in permanent infrastructure to help and net a whole group of plants at once. I am happy for birds to take a proportion of my crops, but there are some species where if you want any at all there is no choice but to net. In my experience, these are currants and black chokeberry, with blueberries and juneberries close behind.

Stop that plant!

Finally, thought needs to be given to how to control crop plants with a tendency to spread beyond their initial planting place. Plants can increase by seed, by division of bulbs, by spreading rhizome or by overground runners. Often this is a good thing, bulking up our supplies of that plant (and replacing what we have eaten), but sometimes a plant can try to occupy more space than we wish to give it. There is a range of strategies that you can use to make sure that you don't spend all your time battling with expansionist plants.

Containers will confine plants that spread aggressively at the roots, like mint and dittander, but plants that make runners, like wild strawberries and silverweed, will throw down their runners and be off, like Rapunzel escaping from the tower. **Soil barriers**, which can be bought in garden centres, are often used to contain large, spready plants like bamboo. As discussed earlier, wide **paths** make good weed barriers. So too can some plants: **a swathe of spring ephemerals** like wild garlic will resist colonisation while they are growing and can be hoed over once they have died down.

You can pit the most spready plants against each other by grouping them together in what I call a **thug bed**, where they can fight each other to a standstill. This kind of bed is likely to require some intervention on the side of plants that look like they are losing the fight if you want to maintain its diversity, but that is a lot easier than trying to control a spready plant in every bed.

Another approach is to **weaken** aggressive plants, by planting them in sub-optimal conditions (such as in more shade than they would prefer) and/or cropping them for green manure. Horseradish, which has a reputation as an unstoppable spreader, is kept in check in my garden like this. In general, plants with large leaves or stems are easiest to control by cutting a proportion of their growth. Japanese butterbur is an example of a plant that spreads strongly by rhizomes but is quite simple to pull out where you don't want it due to the size of its leaves.

Finally, you can simply **avoid** the weediest crops. I have included some quite difficult customers, such as ground elder and nettle, in the crop chapters, but even I draw the line at some, such as sweet woodruff (*Galium odoratum*), where the yield just doesn't justify the work. This line is a personal one and you might draw it in a different place to me.

Whatever the method, it's a good idea to keep spready plants **away from the boundaries** of your plot in order to be a good neighbour (see Chapter 3). You don't want your crops to become your neighbour's weed problem. This also applies when your growing space is adjacent to wild land which might be invaded by your crops.

Managing plant spread is a matter of maintenance as well as design. To prevent too much spread by seed, it is best to **remove the seed heads** from plants like Turkish rocket and sweet cicely before the seed is mature. I usually allow these to flower for the benefit of the insects but remove the heads and compost them once the seeds start to form.

Purple angelica flower,
Royal Botanic Garden Edinburgh

3

Creating Your
Forest Garden

Practicalities to consider

The thing I am most often asked about home gardens is 'where do I start'? The previous two chapters should have given an idea of what you need to consider about your existing site before starting, and how a finished home garden looks and works. The next question is how to get from here to there.

In the idealised picture of a forest garden, you plan everything on a map right from the start, acquire all the plants, prepare the soil and plant everything out into its final position at the same time, generally with large amounts of mulch to fill the empty spaces between the plants. There are several factors that complicate this idealised picture.

- Acquiring all the plants at the same time can only be achieved by buying them all, which is expensive.
- Even if money is no object, forest garden plants can be difficult to source. In reality you will build up your collection over time.
- Most forest garden plants take more than one year – sometimes many more – to grow to their final size, so this approach wastes a lot of growing space.
- Mulch can be difficult to acquire, and it can have a high embodied energy cost and environmental impact.

- Ground preparation is hard work. Going ahead and planting without ground preparation leads to even harder work later.
- There may be an existing garden that you want to transition rather than a blank slate.
- You will undoubtedly change your mind about some things and want to tweak your design during the implementation of your home garden.

The key to the practical implementation of a forest garden is to realise that ground layer plants can generally be moved, but trees and large shrubs can't. The different routes to planting one therefore all involve planting the trees and large shrubs into their final positions at the start, followed by rolling out the smaller plants in a variety of ways that depend on your starting point, which may be uncultivated ground, cultivated ground or an existing garden.

If planting into existing cultivated ground used for annual crops, such as an allotment or a ploughed field, plant home garden plants into their final positions but continue to grow annual crops in between them for the first few years. This is known as catch cropping. It means that you don't waste growing space while the forest garden plants are growing to their full size.

If planting into an existing garden, replace existing plants with forest garden plants as you get them.

If clearing uncultivated ground, don't let your planting get ahead of your ground preparation or your ground preparation get too far ahead of what you can manage to plant up. Plant large, immovable plants into their final positions and mulch around them. Make a nursery bed for smaller plants: for sowing seeds, growing on baby plants and acting as a holding area while you prepare more areas. Start cultivating ground and plant the nursery bed plants out into their final positions as you make space for them and as the trees and shrubs modify ground conditions.

If introducing a new ground layer species later on, try planting several individual plants into different areas and seeing where it does best.

Preparing the ground

Ground preparation aims to remove weeds (including the roots of ones like dock or couch grass which will grow anew if the top growth is removed) and improve drainage and rooting depth. You can also take the opportunity to incorporate nutrients and organic matter into the soil to speed up plant establishment. Methods include hand-digging, rotavation and mulching. Each has its benefits and drawbacks. Remember that with a forest garden you will only have to do this once. If you want to make any major soil alterations, this is the time to get in and do it.

Curly-leaved parsley amongst miners' lettuce, kale, sea beet, strawberry and snowbell

The simplest and easiest method of **hand-digging** is just to loosen the soil with a fork or spade, removing weed roots as you do and spreading compost over the top. If you want more instant results you can 'turn' the soil, turning over a spadeful at a time so that the weeds are buried and the clean, friable soil is at the top. Spreading compost over the top of the soil before doing this means that it gets incorporated into it as you go. The advantage of turning the soil is that you can hoe the soil to a fine tilth and sow seeds into it straight away, and that weeds on top of the soil get composted underneath it, adding to the organic matter. The disadvantage is that you disrupt existing soil flora and fauna more than by just weeding and loosening the soil.

The disadvantage of hand-digging in general is that it is hard work. The advantage is that by digging down into it you get to know your soil better – and you get some exercise! Both the advantages and the disadvantages of digging go double for the method called 'double digging'. This involves digging out a trench to the depth of the spade's blade, loosening the soil below with a fork (or, in extremes, a pickaxe and pinch bar) and spreading a layer of compost at the bottom. The process is then repeated with a parallel trench, turning the soil from the second trench into the first… and so on. At the end, the soil from the first trench is used to fill in the last. While this is a lot of work it is the only way to deal with structural drainage problems and increase the effective rooting depth without large or specialist machinery. Over time I have double dug most of my plot.

There are machines called **rotavators** that mechanise the turning and loosening of the soil. Their advantage is that you can do a large area faster than by hand-digging. Their disadvantages are that they take a fair bit of strength to use and they only churn up a fairly shallow surface layer of the soil. They are suitable for dealing with large areas when you want to create a tilth and there is no need to loosen the subsoil.

Alternatively, you might want to take a no-dig approach from the start, which has the advantage of leaving the soil layers and their associated life forms intact. This is sometimes achieved by **spraying**, which has the advantage of being fast. On the other hand, buying herbicide can involve supporting some very unethical companies. You do not need herbicides for the ongoing maintenance of a home garden and it is unlikely to be worth getting the training and specialist equipment that you need to do spraying safely for a one-off application. There are now alternatives to herbicides that involve spraying plants with hot water, but again these require special equipment that is not worth acquiring for a small job.

Another no-dig method is to lay a **sheet mulch**. A mulch physically prevents weed from growing up through it and excludes light, so that the plants below it die and compost. Woven plastic fabric is very effective and widely available. Cardboard works well too, but you need to lay several layers and overlap any joints by at least 15cm (6in) or weeds will simply push up between them. A vogue for using old carpets as a sheet mulch has thankfully passed. Weeds grow up through them and they become incorporated into the soil. I spent days digging out buried carpets when I first got one allotment.

The advantage of sheet mulch is that it is relatively little work (but don't underestimate how long it takes to lay cardboard properly if you have to remove plastic tape and suchlike from it first). A disadvantage is that it is quite resource intensive, either using fossil fuels for the plastic or diverting cardboard from recycling and increasing the demand for virgin pulp. The other disadvantage is that it is a slow method. Sheet mulch really needs to stay in place for a whole growing season if it is to be effective. Over winter it achieves little as plants need to be growing to be killed by it. Sheet mulching is most effective where you have a large area that you can afford to clear bit-by-bit over the course of a number of years. In this case it is worth investing in a large sheet of woven plastic and moving it on each year, letting it do your work for you in its new place while you plant up the old one. You can try to have the best of both worlds by making holes in the mulch and planting large crops like potatoes or squash through it, but bear in mind that this will reduce the effectiveness of the mulch as a weed killer. In my opinion it is worth doing with cardboard but not worth ruining a plastic sheet for.

Sourcing forest garden plants

In my ideal future, you will be able to get your forest garden plants by going to the forest garden section at the local garden centre or, better still, digging up what you need from your neighbour's home garden. For now though, one of the biggest challenges in forest gardening is getting hold of some of the plants in the first place. You will probably need to resort to a combination of specialist nurseries, interest groups, wild collection and occasionally knocking on a stranger's door and asking if you can have a cutting of that plant growing in their garden. You will certainly need to become adept at raising plants from seed and from cuttings rather than buying them all in convenient pots.

Fortunately, a few nurseries specifically dedicated to forest gardening do exist, such as Martin Crawford's Agroforestry Research Trust (ART) in England or Plants With Purpose and the Red Shed Catalogue in Scotland. Nurseries that sell a wide range of plants can be worth checking out even if home garden plants are not their main focus; Crûg Farm in Wales is a good example. When it comes to fruit and nut trees and traditional soft fruit there is much more choice, and local nurseries can often give advice and supply plants suited to your area. There isn't space to list suppliers for all parts of the world here, and unfortunately nurseries do come and go, but I keep an up-to-date list of ones supplying Europe and North America on my website.

For individual species (in the UK) the RHS's online Plant Finder is invaluable, with reasonably current lists of where you can find almost any plant that's out there.

When buying live plants you normally need to buy from within your own country or trading bloc. Seed lists allow you to be more international, and the ART delivers throughout the EU. B&T World Seeds in France has a very full list and allows you to make requests. Search engines and online marketplaces like eBay can turn up a host of small suppliers for unusual species, often with people offering small amounts of seed from their garden. Seed saver groups like the Scottish Rock Garden Club and KVANN, the Norwegian seed savers' network, pool and distribute seeds from thousands of keen amateurs, and a lot of seed swapping and sharing goes on via Facebook groups, bulletin boards and small websites like my own. Botanic gardens may maintain seed lists for Friends groups and some gene banks, such as IPK Gatersleben in Germany, allow requests from individuals.

Many home garden species are unimproved so it can be worth collecting them from the wild. In Scotland it is legal to collect seeds and cuttings for your own use from the wild but you need the landowner's permission to transplant whole plants. Rules vary, so check local laws in other countries. You can be sure that wild-collected seeds will be locally adapted, while bought seeds may be from a different part of that plant's range.

Home garden plants can be expensive, but they have the advantage over conventional crops that you only ever need to buy them once.

Arranging and keeping track of plants

Once you have your plants, you need to arrange them within the beds. Although it might not look so natural, the best way to do this is to plant in straight lines. Rows give several benefits. One is that you can hoe between them easily, making weeding faster and simpler and helping to prevent spreading plants from taking over the entire bed. It also makes it straightforward to lay plant matter down as mulch between your plants. A lot of waste material in a home garden is quite long: easy to lay along a linear gap but difficult to thread between randomly scattered plants. Finally, many forest garden plants die down and become completely invisible for a good portion of the year. Rows make it much easier to keep track of where you have put something and to avoid planting another plant on top of it!

However, a row of diverse home garden plants is not so simple a thing as a row of lettuces in an annual garden. My standard row is 50cm (20in) wide, but some plants are wider than this and some are smaller. The sketch (right) shows how these can be arranged into useful rows all the same. Smaller plants are planted in patches the width of the row, while larger ones are planted in the middle of the row and overhang adjacent rows. This only works for single-stemmed plants: large clumping plants must simply be given the space that they need elsewhere. Self-seeding plants don't, of course, care about your tidy rows and will come up anywhere, but self-seeders often need to be thinned out to prevent having masses of tiny, unuseable plants, so don't be afraid to hoe a line through a patch of them.

I also find it useful to mark the position of herbaceous perennials with a stick once they start to die down. Once the garden becomes more established it is simpler to reverse this: only marking spaces where a plant has died or been removed. It is remarkably easy to lose track of these new planting opportunities if you don't.

Above all, when arranging your plants, don't be tempted to plant them closer than the space they will eventually require. This is a bit of a 'do as I say, not as I do' rule, as I have often succumbed to the temptation to squeeze another plant into a space that doesn't really exist, but I can say from

Arranging diverse plants in rows

experience that it always leads to more work further down the line. (At least I am in good company: apparently Robert Hart's pioneering forest garden was notoriously overcrowded.) Use catch-cropping to make use of the space left between plants as they grow to full size instead.

Introducing a new plant to the garden

The process of acquiring new plants is usually a continuous one, so you will often find yourself with a new plant to try to fit into your system. Where you have a single plant this is simply a case of reading up on its requirements and trying to find the spot that fulfils them best. Often, however, you will have more plants than you will eventually need, raised from seeds, divisions or cuttings. Even when buying plants you can often get extras much more cheaply than the first plant. You can use this to allow the plant to find its own place in the garden, which in my experience doesn't always correspond exactly to what the books say. Plant your new plants out in a range of positions, varying in degree of shade and drainage. Later on you can take out the ones that don't do so well, ready for the next experiment.

Planting trees in the community

Keeping the neighbours happy

No garden is an island, and being a good neighbour is always valuable – but especially if you are trying out a new and poorly understood method of gardening that can look to the untrained eye like an uncontrolled mess. This is perhaps most true on allotment sites, with rules, committees and neighbours on all sides, but even if you live somewhere where your only neighbours are owls and foxes, your garden will still have an impact beyond its borders.

Most obviously, your garden should not have a negative impact on surrounding ones. This means not shading them unreasonably and keeping aggressively spreading plants back from your borders and under control. If anything, this applies with even more force if your garden borders onto wild nature. In Scotland, the Wildlife and Natural Environment Act (WANE to its friends) prohibits the release of non-native plants into the wild, and you don't want to be responsible for introducing the next rhododendron or Japanese knotweed. In this situation, think carefully about the potential of any new introduction to spread by seed: plants spread by the wind and by birds are especially likely to become a problem.

A neighbour with the Japanese plum harvest

On allotment sites, it is your responsibility to ensure that both neighbours and officials understand what you are doing and that your plot is well managed despite not looking like a traditional one. Keep your plot tidy and use straight lines where you can. Find out the allotment rules on things like tree heights and stick to them. Labelling plants clearly is useful in its own right, but also helps others to understand that a plant is a crop and not a weed.

Even consider putting up signs explaining the techniques that you are using and the uses of your crops. Having a nosy around other people's plots is an ineradicable allotment tradition and you might be surprised at how many people you reach!

Above all, take the time to talk to your allotment neighbours and officials. On the small site where I have my home garden we all have very different gardening styles but we get on well, take an interest in each other's plots and share produce. One June I received a letter from Aberdeen City Council telling me that my plot was not cultivated and I must dig it over or leave. The email that I sent them in response ran to six lines of crops that I had already harvested that year and six more of ones that I still had to harvest. I invited the assessor to come and let me explain how I used the different crops, and to her credit she took me up on the offer. After half an hour of detailed descriptions of how I cooked and ate each weedy-looking plant the poor woman took the point and I have been allowed to continue with my odd ways ever since.

There may be other opportunities to become an ambassador for home gardening. I always try to say yes to requests to show people around my allotment and the perennial-based community gardening projects I am involved in. Some allotment associations, neighbourhoods or villages have garden open days that you can take part in. The council housing estate where I stay takes part in the Beautiful Scotland competition each year. Although I was sceptical about this at first it has increased pride in the area, outside recognition and council assistance in a virtuous circle. Fortunately Beautiful Scotland has gone beyond the days when they used to count bedding plants for their awards and they now recognise factors like community involvement and wildlife value. My allotment is now an established part of the tour for assessors, which led eventually to an award from the Royal Caledonian Horticultural Society (better known as the Cally) for services to horticulture. All these things help to raise the profile and respectability of home gardening.

Compost bin fauna

4

Managing Your Forest Garden

Once a home garden is established, it needs to be maintained. Weeds and pests need to be managed, harvests need to be picked and the nutrients that you take out as crops need to be replaced somehow. Pruning your plants can increase yields and keep them accessible. Plant breeding can improve both the quality and the quantity of your harvests. This is not the 'no work' version of forest gardening that is sometimes promoted, but I suspect that I wouldn't have it otherwise. These tasks draw you into engaging more deeply with your edible ecosystem, and if I didn't have to do them I would have to find some other excuse to be pottering around in the garden.

Feeding the garden

A forest garden excels at hanging onto nutrients, but it still needs to get those nutrients from somewhere. A few will be weathered from the bedrock or come down in the rain, but if you keep taking fertility out in the form of crops then inevitably you will have to put some back in. Maintaining garden fertility means both being careful not to let any garden 'waste' go to waste and topping up the nutrients in the system from time to time with external inputs. Of course, you can always have a productive soil by banging in lots of these inputs, but many of them, such as artificial fertilisers, have a high environmental impact or cost lots of money.

Assuming that sustainability is part of the reason why you are doing forest gardening, the challenge is to obtain these extra nutrients from low-, no- or positive-impact sources.

An understanding of the different plant nutrients helps you to get them to the right places at the right times for your plants.

Plants use light to combine **carbon** dioxide and water from the air and soil to make cellulose for their cell walls, lignin to stiffen stems, sugar to barter with animals and fungi, and a host of other compounds. As handy side-effects they produce oxygen and remove greenhouse gases from the atmosphere, preventing the Earth from suffering the same runaway greenhouse effect as Venus. Carbon compounds also play a key role in soils, altering soil texture and holding onto water and nutrients. Fortunately there is never any need to add carbon as a nutrient, since it is freely available from the atmosphere.

If carbon is about structure, **nitrogen** (chemical symbol N) is about function. It is a major component of the bases used to store information in DNA and the proteins that choreograph reactions in cells. It is essential for chlorophyll, the green pigment that plants use to capture light in photosynthesis. A lack of it results in stunted, yellowed plants. Gaseous nitrogen (N_2) is the largest component of the atmosphere but few plants are able to use it directly. They need it to be combined with other elements to make less stable forms such as nitrate (NO_3) and ammonium (NH_4) compounds. These are highly soluble so they are at constant risk of being washed out of the soil. They are found in inorganic fertilisers created with fossil energy and in compost and animal wastes such as manure.

Phosphorus (P) is used heavily by animals as well as plants. We use it to build our bones and grow our teeth. Large parts of the Scottish landscape are phosphorus-limited, which is to say that it is the scarcest element relative to plants' needs. In the Pacific North West of North America, phosphorus brought from the sea by salmon returning to spawn is the basis of one of the richest forest ecosystems on earth, but in Scotland this process has been interrupted. The vegetation around abandoned dwellings is often visibly richer than the surrounding landscape, especially around the outside toilet! In plants, phosphorus plays a wide range of roles in cell function. Deficiency can show up in slow-growing plants with darkened, purple leaves (although other things can also cause these symptoms). Sources of phosphorus can be animal (bone meal, guano or manure), vegetable (compost) or mineral (ground-up phosphate rocks).

Potassium (K) plays a role in disease resistance and in fruit and tuber production. Deficient plants can show yellowing in between the leaf veins, as with nitrogen, but there is a way to tell the difference. Nitrogen is quite mobile in the plant and is moved to where it is needed, so the older leaves turn yellow as the plant prioritises new growth. Potassium is less mobile, so it is the new growth that suffers as supplies run out. Potassium deficiency can also cause necrosis (dead, brown patches) around the edge of the leaf. The name of the element is derived from 'potash', the traditional source of it. Plant material such as wood or ferns was burned and the ashes were soaked in a pot of water to extract the potassium. Wood ash still makes a good fertiliser, although it may contain large amounts of calcium and sodium as well. Being highly soluble, potassium is easily leached

'Chlorotic' leaves indicating a lack of nitrogen. Leaf veins remain green while the areas between them turn yellow.

Phosphorus-deficient maize

out of sandy soils. The solution to this is, of course, to grow a forest garden and retain the potassium in the living system!

Together, nitrogen, phosphorus and potassium (NPK) are known as the macronutrients. There are also **micronutrients** such as calcium and iron, but these are unlikely to be an issue unless you live in an area with a deficiency or rely too much for too

Bean leaf showing signs of potassium deficiency

long on inorganic fertilisers. Extremes of pH can also cause some nutrients to become bound up and unavailable to plants.

Plants vary in how much feeding they need. As a rule of thumb, the more you take out the more you need to put in. A plant where you harvest leaves or flowers at a low intensity can probably get enough from the rain and the soil. Root crops and intensively-harvested leaves need extra nitrogen, while fruit crops can use extra potassium.

Composting garden material

Many gardens export a shocking amount of fertility by binning and burning biomass. You should practically never need to do either of these. However, garden waste does need to be broken down before it can be re-used by your plants, a process better known as composting.

Compost heaps are good for cycling fresh, non-woody material. I use plastic 'dalek' style bins in the middle of my beds so that nutrients lost out of the bottom of the heap are taken into the surrounding bed. Fresh material goes in the top and compost is taken out of the bottom, either via a hatch in the

fancier ones or by occasionally pushing them over with the more basic ones. This continuous process allows the bin to become colonised by a full complement of organisms that help the breakdown of the material, such as soil microfauna, worms and snails (see p.62). This, and the lack of the dry edges found in a more open heap, mean that there is no tedious turning of the compost to be done.

Compost heaps allow you to move fertility around the garden, and mine partly take fertility from deep-rooted perennial plants to the more nutrient-hungry annual beds and patches. The downside of this is the possibility of moving weeds and plant diseases around too. Closed compost heaps also score here because the exclusion of light kills most plants, but beware of bulbs, tubers and other persistent plant parts. I keep a separate 'toxics' heap for these where they get a longer composting time that makes sure that they are dead. I also use this bin for things like brassica roots and diseased potatoes that might spread disease to subsequent crops, but on the whole you don't need to be too worried about diseased plant material. Most plant diseases, like rust or mildew, won't make it through the composting process or are ubiquitous in the soil and air anyway.

The most important rule in composting is to

Nutrients from the compost bin feed the surrounding plants

Waste material laid between rows as a mulch

maintain a balance between 'green' and 'brown' materials. Green materials are those that are high in nitrogen and tend to form a slimy, evil-smelling gunk if you pile up too many of them. This is because they break down and lose structure quickly, stopping air getting in and leading to anaerobic decomposition. Kitchen waste and grass clippings are examples.

Brown material is drier, more fibrous and higher in carbon. It helps to keep a heap open and aerated, but breaks down much more slowly. The result is that if you have too much of it the composting process will stop entirely, meaning that you are only storing, not composting, your waste. Brown materials include long grass, dry plant stems, autumn leaves and anything woody.

General garden weeds are usually towards the middle of the green-brown spectrum. They will help the composting of either extreme and don't need any special treatment.

Getting the balance between green and brown

materials in a compost heap right is a matter of experience. Things go best if different kinds of material are mixed together well. In a traditional compost heap this is achieved by layering brown and green material in thin alternating layers and laboriously 'turning' the heap from time to time, but in a forest garden it mostly happens fairly naturally as you rarely have large amounts of one kind of material going in at once in any case. Turning the heap is not a task that I miss.

In a mature home garden you will have a surplus of brown material, not all of which can go in the compost. A lot of it will be in the form of long stems which are difficult to fit into a compost bin anyway. **Burning** this does give you potassium-rich fertiliser to use on your fruiting plants, but it also sends all the hard-won nitrogen up in smoke and wastes a source of soil organic material.

Options for dealing with it include **leaving it in situ**. This may be the best for wildlife such as the invertebrates that like to overwinter in hollow

Path mulched with hedge clippings

stems, but can lead to the transmission of diseases like apple scab, which overwinters in apple leaves on the ground, or may simply get in the way. I remove apple leaves and compost them along with richer material in my compost bins.

The next option is to use it as a **mulch**, by laying it on the ground around larger plants that don't have much growing under them or between the rows of smaller plants. The key to breaking down this kind of material is that it takes more time, so putting it where it takes up otherwise unused space helps. It also needs to be kept moist and close to the soil; material left 'high and dry' by piling it too high won't break down. You can spend money on a shredder to speed things up and create a more manageable product, but I have never felt the need and you would be reducing the wildlife value of the garden. Woody material should never be dug into the soil as bacteria then pull nitrogen out of the earth to help them digest the carbon.

With the woodiest material, such as old raspberry canes, I turn the problem into a solution by laying them on **paths**. Other options for using wood are on p.74.

Plants for fertility

Many books on forest gardening suggest planting some plants specifically for fertility. I have slowly done less and less of this as I find that the crop plants fulfil all the roles usually suggested for them, but you might want to use some. There are three main classes of plants that are suggested for forest gardens in order to provide fertility: deep rooters, nitrogen-fixers and 'dynamic accumulators'. In organic gardening, plants that are sown in one year in order to dig into the soil for organic matter the next year are known as 'green manures'.

Most of the action in soils takes place in the top 30cm (12in). Rain and organic matter fall on the surface. The organic matter is broken down and absorbed into the upper layers of the soil.

Sweet cicely root

Most plant roots are in this layer, attempting to get first dibs on the water and nutrients coming from the surface. Others, however, go much deeper.

Some plants do this for stability, to get a better hold on the soil. Others are probably taking up nutrients from the lower levels of the soil, where they weather out of the base material or get washed from upper levels by the rain. Others, with thick, fleshy tap roots, may be taking nutrients in the other direction, using the lower soil horizons as a pantry, storing nutrients safely away from frost and gnawing animals.

Some forest gardeners plant **deep-rooted** herbaceous perennials specifically to bring nutrients to the surface, comfrey (*Symphytum officinale* or x *uplandicum*) in particular. Comfrey is deep rooted and a fast grower, outcompeting weeds and producing several large harvests a year of material that can be composted and fed to other plants. Plants like comfrey can be seen as perennial green

manures, taking up part of the garden in space instead of time to provide organic matter and nutrients for other plants.

I question, however, whether any plants are required that fulfil this function only and whether it is a wise use of space in a small garden. Many crop plants, such as most of the trees, sweet cicely (and lots of other members of the celery family), horseradish, Turkish rocket, herb patience and udo, provide the same service while also providing useful products. I once tried to dig up a sweet cicely root whole. At a metre's depth it snapped off, but since it was still several centimetres thick at that point I assume it went considerably deeper. While digging out deep carbon paths (p.27), I have found roots of apple, horseradish and Turkish rocket down to this depth and more too.

I call these deep-rooted crops 'dual use' plants, in that they act both as a crop in their own right and as a green manure. Many of them either shed leaves or require cutting back at some point in the year, either to generate new shoots, as with sweet cicely, or to restrain their growth, as with horseradish. The material generated in this way can be composted by any of the means above.

Nitrogen-fixers are a popular choice as green manures, since they have the unusual ability to make useful nitrogen compounds directly out of atmospheric nitrogen instead of scavenging them from the soil like most plants. They do this through a symbiosis with nitrogen-fixing bacteria that live in their roots and are rewarded with sugar made by the plant. Most are in the pea family (Fabaceae), but a scattering of plants from other families have also mastered the trick. These include the alders (*Alnus*), bayberries (*Myrica*), silverberries (*Elaeagnus*) and buffalo berries (*Shepherdia*).

There is less opportunity here to grow dual-use plants, since useful perennial crops that also fix nitrogen are rare in cool temperate areas. With those that do exist, I regard their nitrogen-fixing abilities as a bonus rather than a major part of my strategy for feeding the garden.

You can grow annual nitrogen-fixers in the ordinary way in your annual beds or in gaps in the forest garden. Alternatively you can grow perennials purely for green manure and cut them for material to compost. Graham Bell in the Scottish Borders grows laburnums (*Laburnum anagyroides*) which he coppices and puts in enormous compost heaps that are then planted with squashes. In a small garden, however, the amount of room that you might want to dedicate to such plants is limited, and you might want to put more effort into putting in nitrogen from outside sources (next section).

Finally, forest gardening books sometimes recommend the use of '**dynamic accumulators**', defined as plants which actively concentrate nutrients from their surroundings and hold them in the living system. Since this is practically the definition of a plant, it is difficult to be sure what the term adds, but it is true to say that some plants are richer in particular nutrients than others. The big drawback is that unless a plant is deep-rooted or nitrogen-fixing, the only nutrients that it can accumulate are those from the top layer of soil, so those nutrients that you most want to add are

Flowers of the nitrogen-fixing (and pretty) *Lathyrus tuberosus*

those that you are least likely to get. On the whole, this is another function that you can rely on your crop plants to fulfil. I have written more about the idea of dynamic accumulators and whether it is a useful term on my website.

CROPS THAT FIX NITROGEN

Pea trees (*Caragana* species)

Charma and sea buckthorn (*Hippophae* species)

Soapberry and buffaloberry (*Shepherdia* species)

Bayberries and bog myrtle (*Myrica* species)

Liquorices (*Glycyrrhiza* species)

Alfalfa (*Medicago* species)

Talet (*Amphicarpaea bracteata*)

Hopniss (*Apios americana*)

Aardaker (*Lathyrus tuberosus*)

Milk vetches (*Astragalus* species)

Cairmeal (*Lathyrus linifolius*)

Broad beans (*Vicia faba*)

Runner beans (*Phaseolus vulgaris*)

Peas (*Pisum sativum*)

Other nitrogen-fixing plants

Alders (*Alnus* species) – trees

Laburnums (*Laburnum* species) – trees

Furze/whin (*Ulex europaea*) – shrub

Broom (*Cytisus scoparius*) – shrub

Lupins (*Lupinus* species) – annual/herbaceous

Clovers (*Trifolium* species) – annual/herbaceous

Fenugreek (*Trigonella foenum-graecum*) – annual

Winter tares (*Vicia sativa*) – annual

Black medick (*Medicago lupulina*) – herb

Birdsfoot trefoil (*Lotus corniculatus*) – herb

External inputs

You can also put fertility into the garden from external sources. Home gardeners generally try to avoid fertilisers derived from fossil fuels or heavy dependence on external inputs for environmental reasons, but some inputs have a lower environmental impact or would otherwise become waste. Because the forest garden is good at holding onto nutrients, you can be quite opportunistic, adding materials as and when they become available rather than sticking to the strict schedule of applications needed by the annual garden.

Composting **kitchen waste** for your garden turns a waste problem into a valuable resource. There are a lot of myths out there about what you shouldn't put into compost; you might have heard stern warnings against adding citrus peel, rhubarb leaves, coffee grounds, rice, eggshells, onions and garlic and many more. These are all nonsense. Some materials might take longer than others to break down, but we are in no rush and have no need of making perfect, peat-like compost. Most garden compost heaps don't achieve the holy grail of 'hot composting' where the pile gets hot and breaks down very quickly anyway. Leaving out lots of material just makes it all the more likely that your heap won't reach the critical mass of material needed.

Meat, dairy, bread, fat and other highly digestible foods do present a challenge. If you do put them in, your compost heap will love them, but there is a risk of attracting rats. The easiest way to solve this is to run a thrifty kitchen that doesn't produce such waste in the first place. Soup is better than compost!

Kitchen waste tends to be 'green' in composting terms: it becomes sticky and smelly if piled up on its own. This makes it a good complement to garden waste and both go in the same bin.

Before kitchen waste goes to the garden it must be collected in the kitchen. Many people try to hide away and cover their compost collector. I take the opposite approach. My compost bucket (which is literally a bucket) has pride of place in my kitchen, so it is easily accessible. It is open: closed collectors become humid, promoting smelly, anaerobic decomposition and providing an ideal home for fruit flies. It is emptied regularly and when I begin a new bucket I line it with a few sheets of paper. This absorbs any liquid coming out of the compost and stops it becoming smelly. Be proud of your composting!

Some other **household waste** can also be composted if it is of plant or animal origin. Dog and cat waste pose too high a risk of disease to go in, but droppings or bedding from herbivorous pets like rabbits are fine. Sweepings and the contents of the vacuum cleaner are mostly human hair and skin flakes (sorry!) and will break down nicely. Paper and cardboard are better recycled, but soiled paper such as chip wrappers will disrupt the recycling process: I save them for lining the compost bin. Paper is very much at the 'brown' end of the compost spectrum so too much of it is a bad idea.

In a modern house, the risk with home composting is introducing contamination into your bin, your soil and your food. While this is on one hand a bad thing, the knowledge that your waste is going into your garden is a good consciousness-raiser, making you think more carefully about what you use in your household in the first place.

The biggest concern is plastics, which now crop up in the oddest places, from chewing gum to teabags. Larger lumps of plastic in your soil are unsightly, but the bigger worry is the microplastic particles into which they inevitably break down. The effect of these on human health is a great unknown at present, but they have been shown to be harmful to wildlife and it is surely sensible to avoid them where you can. Microplastics are now ubiquitous in air, soil and water so they can't be avoided completely, but you can take reasonable precautions against adding to the load in your garden.

Obviously, don't put plastic in your compost bin. This includes the many paper/plastic laminates such as paper cups and a lot of wrappings. Some

Food for the garden; wood ash; lovely leaf mould

teabags contain plastic: look for a plastic-free brand or use loose tea in a strainer that goes in the cup; this will save you money anyway. Don't compost your hooverings if you have a carpet made of artificial fibres. Try to avoid clothing that sheds plastic microfibres. Washing machine lint used to be fine to compost but nowadays it usually contains plastic. Beware of 'compostable' plastic bags: often they will only break down fully in an industrial composter.

If you're really serious about creating a closed loop for nutrients in your garden, then the other obvious things to recycle are **urine and faeces**. Human faeces – or 'humanure' if you prefer – is chock-full of pathogens and smelly to boot, so it shouldn't be used in the garden without a lot of care. Composting it is beyond the scope of this book, but there are whole other books on the subject, such as *The Humanure Handbook* by Joseph C. Jenkins. Urine is much less dangerous, although not completely without hazards and shockingly

smelly if stored. It is a perfect nitrogen fertiliser, being full of urea, the nitrogen compound used in many commercial fertilisers. It can be used on crops with a high nitrogen requirement or to add nitrogen to compost bins.

If you have a wood fire then **wood ash** is another waste product you can use in your garden. The chemical composition of wood ash depends on the wood burned, but it is generally a source of calcium, potassium, magnesium and phosphorus. It is a strong alkali, so don't use it on any acid beds and don't overdo it if you have a large fire and a small garden.

Autumn leaves also have to be used with some care. They are very much at the brown end of the compost spectrum. You can use them with care mixed to balance green material in the compost bin or compost them separately over the course of a couple of years to make a peat-like compost called leaf mould, but the simplest way to use them is the way that nature does: spreading them over everything as a winter mulch to add organic matter to the soil. They can be easily available in large quantities due to people's obsession with removing them from where they fall.

In some situations you will have access to **other people's garden waste**. Perhaps someone on your street or allotments bin their grass, leaves or hedge clippings. Should you take advantage of this? In general my reaction is an enthusiastic yes, but with a few cautions. One thing to be careful of is what might be coming along with the waste. In particular, read the section on New Zealand flatworms (p.62) to make sure that they don't come as hitchhikers. Also bear in mind that such material can come in large volumes that are harder to use in a balanced way.

Weeding

A forest garden doesn't need as much weeding as an annual one since the crops occupy more of the space for more of the year, leaving less room for unwanted plants to take advantage of, but it is still a garden, not a truly wild space. As such there are plants that we want and, as a consequence, plants that we don't want – the very definition of a weed. While the conditions in a forest garden make life hard for most familiar weeds, there are also plants that are adapted to invading perennial and shaded plant communities, so over time I'm afraid that any forest garden will accumulate its own set of weeds. Also, your own crops can be 'weeds' where they try to claim more of the garden than you want to give them.

The actual weeds that you have will vary from area to area, but they tend to fall into a small number of classes according to the strategy that they use to spread. The **superfast seeders** are a problem mostly only in the early stages of the forest garden or where you leave spaces for self-seeding crops. They are plants, like the bittercresses (*Cardamine*), and annual meadow grass, that germinate and

produce seed while still small so they are always in the seed bank. It's worth learning to recognise these at their tiniest stages so that you can weed them out of sown rows, but once perennial plants get established they stop being a problem.

The **fleshy-rooted seeders**, such as dock and dandelion, are similar. These are troublesome in the annual garden but you can be relatively relaxed about them in the forest garden, and even think of them as one of the ways in which the garden stores fertility. The strategy of this class is to put lots of stores into their roots, so they are not much worried by having their top growth pulled off or hoed, simply sprouting up again before long from the root. The main thing with these weeds is not to let them seed and, when you get round to removing them, to take out the whole root with a trowel or planting spade.

Another way that plants avoid being weeded out is by having **brittle stems**. Chickweed and stitchwort are examples. Their stems are brittle enough that pulling them never pulls up the whole plant: to get rid of them you have to hoe or to trace the stems back to the base and pull out the roots, which is fiddly.

Such plants can be a major problem in the annual garden (chickweed in particular is the bane of my allotment neighbour's life) but in the perennial garden they are only much of a problem in the establishment phase or when they combine the trait with **scrambling or twining** up other plants. *Galium aparine*, variously known as sticky Willie, goosegrass or cleavers, is one of the worst of these, being adapted to invading perennial communities by germinating early in the year, growing fast and climbing up and over other plants with brittle stems covered in tiny hooks. Bindweed (*Convolvulus arvensis*) is another culprit. It twines around perennial stems and its beautiful morning-glory flowers invite the gardener to leave it a while.* Don't be tempted! This is one class of weeds that I have close to zero tolerance for.

* In the great gardening aphorism coined by my friend Jodie Peacham, 'It's the pretty ones that get you'.

Dandelion flowers

Bindweed flower (on someone else's allotment!)

Sticky Willie

Chickweed has an extra trick up its sleeve: it can produce new roots along its stems, making it even harder to get out without damaging nearby seedlings once it is well established. Plants with **runners** have taken this strategy one step further, with specialist stems dedicated to spreading out and conquering new territory. Because the new plantlets are fed by the parent plant, they can root into the crowns of other plants in a way that no seedling could. Creeping buttercup (*Ranunculus repens*) is the worst example of this class in my garden. It combines this trait with an ability to grow significantly over winter, letting it take over a lot of space when gardeners' backs are turned.

Other plants do the same thing but below ground, with underground runners, which can either be roots or specially adapted stems. These are particularly disruptive in the perennial garden as the best way to remove them is to dig them out, disturbing the perennial plants. Make sure your ground is clear of them before planting and don't let them get established afterwards. Examples include horsetail, silverweed, ground elder, nettle, mints, couch grass, Yorkshire fog (another grass), rosebay willowherb, silverweed and many bamboos. You can also get rid of these weeds by constant removal of the top growth, weakening the root system and eventually killing them, but you need to be persistent and it is much easier with large-stemmed plants than small-stemmed ones.

With all runnering plants above or below ground, the most important thing is to avoid leaving reservoirs that they can continually spread out from, such as beds edged with stones, groups of pots or piles of 'useful gardening materials'.

Hardest of all are plants that combine one or more of the above traits with **shade tolerance**. These are the true woodland weeds, such as ground elder (*Aegopodium podagraria*), woodruff (*Galium odoratum*), wood poppy (*Meconopsis cambrica*) and wood avens (*Geum avens*).

The dividing line between a desirable plant and a weed is a thin and movable one. Many – even most

– of the plants listed above have edible or otherwise useful parts. Some are even listed in the crop chapters later in the book! Some that I class firmly as weeds are sold in nurseries as forest garden plants. Which plants you decide are more trouble than they are worth will depend on how much you like that plant, whether there are better-behaved alternatives, whether the conditions in your garden favour its imperialistic tendencies and whether you are able to forage it outside your garden (lots of weedy plants are, not surprisingly, widespread outside gardens).

Creeping buttercup

Field horsetail's underground runners allow it to form dense stands

This edging looks pretty but is a weed reservoir waiting to happen

Some people use herbicides to deal with weeds. I would discourage this in the forest garden, partly because life is easier if you don't have to store and use poisons, but mostly because they aren't much use. Herbicides are effective at completely clearing an area, which is something you only ever want to do once in a home garden, at the start. They are not easy to target so exactly that you can take the weeds out of a mix with crop plants. The risk of losing your precious crops in order to make weeding slightly faster really isn't worth it.

My main strategies to control weeds are planting in rows that I can hoe between, avoiding weed reservoirs (see also Chapter 2 on paths) and preventing crop plants from seeding when I don't want them to. This is usually easy to do by removing seed heads before they seed, often providing you with something to either compost or eat in the process.

Propagating your plants

Plants can be multiplied in many ways: from seeds, tubers, bulbils and other propagules; by dividing plants; from cuttings; and by layering.

Many plants form spreading clumps by multiplying themselves underground, from roots, rhizomes, runners, tubers, corms and bulbs. Whatever the method, this makes it easy to propagate them simply by **lifting and dividing** the clump. This is best done in winter while the plant is dormant or in early spring when new growth is visible but still small. Often it is easy, but some plants like bamboo need some gentle persuasion. The most effective and least damaging method is to push two garden forks, back-to-back, into the middle of the clump and lever them apart. Sometimes a rhizome has to be actively cut to separate new plants. Where you do this, make sure that there is a new bud on each part that you remove. (This is also true of tubers like potatoes, which can be cut if an 'eye' is left on each part.)

Plants can also make our lives easy by producing **vegetative** propagules (a propagule is any bit of a plant that it uses to propagate itself) above ground, often in the seed head. Some plants, like many in the onion family, produce little bulbs called bulbils which can be planted in spring the same way as regular bulbs. Bulbils may start to grow while still on the parent plant, forming whole baby plants which can simply be removed and potted up until they are large enough to plant out. Other plants, like strawberries, form little plantlets on spreading stems called runners.

Almost as helpful as this are the plants whose above-ground growth retains the ability to form new roots, meaning that small fragments can be turned into new, separate plants. The rooting process can take place while the fragment is still attached to the original plant, in which case it is known as layering, or after it has been removed, when it is known as a cutting.

Some plants, such as brambles and their relatives, layer quite naturally by rooting at their tips, and the challenge is not so much to persuade them to do it as to discourage them from rooting everywhere! Otherwise, **layering** is usually achieved by bending a branch down and pinning it into the soil. This method is usually used for plants whose ability to root is weak, so that the new plant needs to be supplied by the original while it grows its own system. Plants that can be propagated this way include apple rootstocks, hazel, plum, and elder. The alternative is 'air layering', in which a twig is encouraged to produce roots by wrapping it in a water-retentive material so that the bark is kept wet. I have propagated apple this way, since layering apples in the ground can lead to soil-borne pathogens killing the new plant and sometimes the old one. Whatever the method of layering, breaking (but not snapping off) or partially ring-barking the stem to be rooted stresses it and makes roots more likely to form.

Cuttings come in four forms: hardwood, softwood, semi-ripe and root cuttings. **Hardwood** cuttings are from hardened-off or 'ripe' wood taken in the autumn, winter or early spring. The ideal length is 15-30cm (6-12in), although it is possible to root much larger pieces of some species, like willow or kale. To encourage rooting, the lower cut should be made just below a bud, or current year's growth can be removed with a small piece of previous year's growth (known as a heel).

The tip of the cutting should be cut off just above a bud. This prevents the vigorous buds at the end of a shoot from growing too quickly in the spring. It is better to take cuttings from vertical growth; branch wood has sometimes become so committed to the horizontal lifestyle that it will continue growing sideways even after rooting! Willows, elders, currants and gooseberries (*Ribes*), honeyberry, sea buckthorns, Darwin's barberry and viburnums are all suitable for hardwood cuttings.

Autumn or winter harvested cuttings are planted in the soil, with most of the length underground. Spring harvested cuttings can be placed with the bottom third to a half in a container of water. Planted cuttings need good weeding as they take some time to establish. Water-rooted cuttings can be potted up once they begin to develop roots, and grown on before planting out.

It is also possible to root the new, tender growth of some plants, known as **softwood**. Plants again vary in how easily they will form roots. Mint cuttings need nothing more than a few days sitting in a glass of water, while others need rooting hormone, healing, misting, shading and disease control.

A good length for a softwood cutting is 5-10cm (2-4in). They should be taken from non-flowering shoots, which are more likely to root. It is best to cut them in the morning while they contain plenty of water, and in spring or early summer. Like hardwood cuttings they can be cut off below a leaf joint (where the bud will form in mature growth) or with a heel of the previous year's growth. The tip is again removed, plus all but a few small leaves

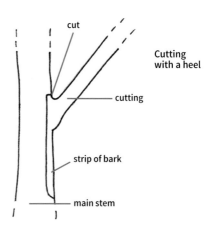

cut

Cutting
with a heel

cutting

strip of bark

main stem

Roots induced to form on a branch of an apple tree by air layering. The branch was then removed and is now an independent tree.

A cutting of the Mexican herb epazote (*Dysphania ambrosioides*) forming roots after two weeks in a jar of water.

towards the top of the cutting. The two biggest risks to a softwood cutting are disease and drying out, so if the cutting can't be potted up immediately it should be stored in a plastic bag in the fridge to keep it fresh.

The bottom of the cutting can be dipped in rooting compound, which usually contains plant hormone to stimulate rooting and fungicide to prevent the cut from becoming diseased. Then fill a pot with potting compost, make a hole in the compost with a finger and insert the cutting. Water from above to settle the compost.

The pot now needs to go somewhere with light but not direct sunlight, such as a north-facing window sill or under fleece. Bottom heat will help to stimulate rooting so a heated propagator will help. Increasing the humidity with a covered propagator or a plastic bag over the top of the pot helps to stem moisture loss, but the cuttings must be ventilated at least twice a week to prevent

disease. The compost should be kept moist until the cuttings start to root. When they do, ventilation should be increased slowly to harden them off.

Softwood cuttings can be used to propagate brassicas, bellflowers, red valerian, fuchsias, mallows, mints, oregano, rosemary and orpine with relative ease and are worth trying with any plant that puts up non-flowering stems with multiple leaves.

Semi-ripe cuttings come somewhere between the softwood and hardwood. They are used mostly to propagate shrubs – often evergreens – such as chokeberries, saltbush, barberries, flowering quinces, salal, bay, honeyberry, bayberries, mulberries, rosemary, sage, bladdernut and blueberry. Cuttings of 10-15cm (4-6in) are taken in high to late summer, when the base of the cutting is hard but the tip is still soft. Unlike hardwood cuttings, small horizontal shoots with short internodes (the distance between buds) are preferred. Again, most of the leaves are removed, leaving a few small ones towards the tip. If the plant has large evergreen leaves, they can be cut in half to reduce transpiration. Thereafter they are treated in the same way as softwood cuttings.

Root cuttings are of course taken in the winter, when the plant is dormant. They are used for plants that do not naturally divide and form clumps, but can be persuaded to grow from sections of root, such as sea kale (where the sections are called 'thongs') and carrot family members like angelica and sweet cicely.

Fruit trees like apple, pear, cherry and plum are often propagated by **grafting**, in which a cutting (called a scion) of a variety that has good fruit but is hard to root is persuaded to join to the roots of another variety (called a rootstock) that roots easily. This is done partly because it is rare to find both these qualities in the same variety, but grafting also has other advantages. Rootstocks that are poor growers can be chosen deliberately to produce a smaller tree for smaller spaces and ease of harvesting. This stresses the plant and makes it more susceptible to diseases, but the trade-off is that stressed plants actually produce more fruit. Grafted plants also start to produce fruit sooner than ones produced by alternative methods.

Trees are usually bought from nurseries pregrafted, but you can buy rootstocks and try it yourself if you want to propagate a new or unknown variety that does well in your area. It is even possible to graft multiple scions onto the same rootstock to make a 'family tree' with different varieties growing on the same plant! Each species has its own set of rootstocks, which are given in the individual plant descriptions (Chapter 7).

In contrast to grafted plants, trees produced by rooted cuttings or layering are known as **own-root** trees. Own-root fruit trees are generally more healthy and robust than grafted trees, but require more pruning or other interventions (Chapter 4) to encourage fruiting and/or keep the tree to a desired size. I think that they are particularly

suitable for community growing projects where the extra resilience can be useful. With apples and pears, once an own-root variety is established it can be propagated from root cuttings. Roots of at least 15cm (6in) (the longer the better), cut during the dormant period, can be planted like a tree but with about 5cm (2in) sticking up out of the ground. They will take longer to produce leaves in the spring than other cuttings. Own-root plum and cherry trees can be propagated by digging up suckers (shoots coming up naturally from the roots at some distance from the main plant) with a length of root and planting them out.

All of the above methods produce a clone: an exact genetic copy of the 'parent' plant. This is often beneficial, where you have a superior variety that you want to preserve. On the other hand, over-reliance on clones reduces genetic diversity and resilience, and clonal lines can accumulate disease problems and mutations. To reshuffle the genetic pack, or if it is impossible to coax a particular species to divide, we need to turn to sexual reproduction and **seeds**.

Some seeds are simple – sow them in the ground in spring and they will grow – but many have built-in mechanisms to stop them from germinating at the wrong time, and to sprout these it is necessary to make sure that they receive the right triggers at the right times. The Plants For A Future database includes a section on propagation for each species, including the treatments needed for seeds.

Most commonly, seeds can need a period of cold, damp weather (called stratification) before they will grow. This prevents them from germinating in the autumn and having to overwinter as small, vulnerable seedlings. A slightly more complicated situation is what is known as 'warm-cold-warm' stratification, in which the seeds need to experience warm, damp conditions (as in late summer) before the cold period.

The simplest way of giving seeds the conditions they need is to obtain them as early as possible after picking, then sow them in autumn, either in containers or in the ground, and let nature take its course. Good row marking and labelling is essential here, with a record of species, source and sowing date. I keep permanent rows and pots, sowing several kinds of seeds in each to save space, as some seeds can take more than one stratification cycle to feel ready to germinate.

If seeds are received too late for outdoor sowing, the necessary conditions can be replicated by storing seeds in the fridge (not freezer), either in tubs of damp sand or on damp tissues sealed in sandwich bags. Warm stratification can be achieved by leaving the tubs somewhere warm for a time before they go in the fridge. At the end of their stratification period the seeds can be sown as normal, but they may also germinate in the fridge so they need to be checked regularly. Clear tubs and bags make this easier.

Plants may also delay the germination of their seeds by wrapping them in a hard, impermeable seed coat, like the skin on a bean. Sprouting of these seeds can be speeded up by removing, nicking or carefully sanding down the seed coat, a process called scarification.

Seed saving and plant breeding

Modern agricultural crops are often practically unrecognisable from their wild ancestors. They have been bred for hundreds, often thousands, of years, for qualities such as yield, size, flavour, disease resistance – even, in the case of the carrot, to be a colour matching the name of the Dutch royal house! By contrast, forest garden crops have often undergone no modification whatsoever and are used in their wild state.

This is a limitation, but also quite exciting, as it means that even amateur growers can make improvements to the genetics of forest garden crops, either by looking out for useful traits in wild or garden plants or by steering the gene pool of the plants in your garden. Obviously, this is easiest with plants that seed every year or two, and rather harder with plants like those bamboos that set seed once a century!

What you are unlikely to want to do is to create a crop variety in the plant breeder's traditional sense. This is a population of plants with very uniform traits. Breeders must keep these populations in large numbers (to ensure sufficient genetic diversity within the population), isolate them from others of the same species and ruthlessly breed out traits that don't match the variety description. This kind of breeding is best left to the professionals.

We are likely to want the opposite: a population of plants with plenty of diversity, well adapted to our own site and uses. Most seed lines are bred neither for forest gardening nor for cold climates, so the genetic base of any crop is unlikely to be perfectly suited to your garden. You can't breed from nothing, so the best starting point is as wide a diversity as possible. I usually try to get several varieties of any species, from several sources.

There are then two possible aims of breeding. One is to combine all the best traits in one plant. You might have one plant that has large leaves (or fruits, roots or whatever) but a bitter taste, and another that tastes great but has small, fiddly leaves. With enough stirring of the gene pool you will eventually end up with at least one individual that has large, tasty leaves. This goal is only worth pursuing, however, if there is a way to propagate this favoured plant vegetatively. Fortunately, a lot of species do combine regular seed production with possibilities for asexual reproduction, including perennial kales, tuber crops, lots of members of the carrot and aralia families (such as skirret, udo and sweet cicely), most members of the onion family and most fruits.

The other goal is to shift the average of the whole population in a favoured direction, for instance towards greater sweetness, spiciness or tenderness, larger size, smaller size, faster growth, fewer spines

– the choice is yours. You can apply both negative and positive selection pressure. It is often necessary to thin out self-seeding plants, so take the opportunity to take out plants with undesirable traits, leaving the best. As an example, I am selecting my population of radishes for better seed pods. Some radishes have rough, spiny pods while others have smooth. Having allowed a few generations for traits to mix thoroughly, I pull out any that show any roughness (if you start this too early, while different traits are still linked within a variety, you may simply end up pulling up all of one variety and lose diversity in traits that you didn't mean to select).

At seed collection time, you can apply positive selection pressure, limiting your collection to the best plants (while being careful not to collect from so few that you narrow your diversity overall). It is often best to mark these plants in advance as their advantages are not always still evident at seed time. Coming back to the radishes, I also select for size, and it is easy to simply collect the largest pods. I have increased pod size several times over the years in this way.

One drawback with some crops is the need to leave the best of your harvest for seeding. In my garden I dig up all the self-seeded parsnips each year, but save and replant a few of the largest, best-shaped and most disease-free to go on and seed the next year. Fortunately this still leaves most of the crop for eating.

Many traits that you will be interested in concern adaptation to your particular garden, such as tolerance of cold, shade or diseases. Fortunately nature will help with this, as plants that do better will produce more seed and contribute more of their genes to the next generation. Over time you may well notice that a population of plants does better in your garden than when you first introduced it. An example in my garden is the leaf beet, which overwinters far better now than when I first sowed it, even taking account of the changing climate. A diverse population adapted over many generations to a particular area is known as a 'landrace'.

HOW MANY SEED PLANTS ARE NEEDED?

To maintain the diversity in a variety, seed savers grow a minimum number of plants, often between 20 and 50, and collect seed from all of them. Growing fewer plants or collecting seed from fewer increases the chances of genes being lost through genetic drift. While these numbers are often not practical in a small garden, you can generally get away with fewer plants if you use these techniques.

- Growing a physically diverse population tends to maximise the underlying genetic diversity.

- Open pollination and avoiding isolation bring in gene flow from beyond your garden.

- Self-seeding or saving seeds over a number of years stores diversity over time.

- Swapping seeds with others makes your plants part of a wider gene pool.

Flat dandelion

Upright dandelion

Beyond this, there are a lot of specific traits that you can select for. A common element in the domestication of rosette-forming plants is selection for upright leaf stems. Plants form rosettes by bending their leaves out and over the ground, helping them to suppress competitors, but this leads to soiled leaves that require lots of washing. Populations often vary in how strongly they bend out, so it is an easy trait to select for. The two pictures of dandelions (left) are both wild plants with no selection, showing how different prostrate and upright plants look.

Short internodes (the distance between leaves on the stem) give more compact plants and mean that the plant has more energy for edible parts. Short, blunt roots rather than long, tapering ones are easier to dig up. Many biennial plants have some tendency towards being perennial, which can be encouraged. Spines and other defences are useful on the stem but less so on the edible parts and can be selected against. Defensive compounds causing bitterness or other bad tastes can be reduced. The toughness or tenderness of leaves and stems also affects edibility and varies between plants. Other traits relate to the ease of harvesting or processing, such as the 'freestone' trait in plums, where the stone comes out far more easily as it is not stuck to the flesh. Colour selection is mostly for fun, but can also be related to the nutritional profile of the plants.

One of the greatest joys of amateur plant breeding is its collaborative nature. Everyone wins from sharing seed and many species have enthusiastic internet forums dedicated to their breeding. This ethos has recently been formalised into the Open Source Seed Initiative. Similar to other 'copyleft' initiatives like open source software or the images on Wikimedia Commons, breeders make their varieties available to others on the sole condition that they do not seek to patent or otherwise restrict the use of the seeds or any future varieties developed from them.*

Plants that are usually propagated clonally by cuttings, tubers or other vegetative parts have a

* This has led to the OSSI being described as Linux for Lettuce.

Not a tribble infestation, but leek seed heads drying out

Chinese artichoke flowers

tendency to lose the ability to produce viable seed over time. Examples are perennial leeks, perennial kales, Chinese artichokes and multiplier onions (such as shallots or potato onions). This narrows the gene pool dangerously and efforts are under way to coax lots of these crops back towards sexual reproduction. Some of these are described in the individual species descriptions in the crop chapters.

For more detail on saving your own seeds, see books such as the *Seed Saver's Handbook* by Jeremy Cherfas et al.

Pests and diseases

As with weeds, pests and diseases are things that you can worry about less in a home garden, but not ignore entirely. Ones that affect individual crops are dealt with in the crop chapters, but there are many more general pests and diseases that can be minimised through broad strategies. It is important to remember that the majority of these are a natural part of the ecosystem and that the aim is not to eliminate them, just to stop them from claiming too large a share of your crops. Where possible, in a forest garden, strategies that involve building a more complete ecosystem are to be preferred to those that involve removing parts of it.

Pests and diseases come in many forms. At the larger end, **browsing mammals** like rabbits or deer can decimate a garden. Since they are mammals like us, with similar tastes, they target precisely the more tender parts of the plants that we are also interested in. Rabbits and deer can be fenced out, but **birds** cannot. They have a particular appetite for seeds and fruit (sometimes enraging the gardener by doing nothing more than pecking a rot-inducing hole in each fruit), but will sometimes attack vegetative growth too, especially in winter. **Squirrels** also laugh at fences to raid nut crops.

Smaller still, all sorts of **insects** chew their way through the garden, sometimes spreading diseases in the process. **Slugs and snails** have an essential place in the garden, unless you want to chew up all your rotting vegetation yourself, but you need to protect young plants from them and ensure that they don't reach plague proportions. By contrast, **New Zealand flatworms** should be eliminated if at all possible. They are introduced invertebrates that attack the fertility of the garden by eating and destroying earthworm populations. Smaller yet, a range of **fungi** and **bacteria** inhabit the garden, causing diseases like rust and mildew.

There are five general strategies that you can apply to all of these: barrier methods, attracting predators, ideal homes, garden hygiene and learning to share.

Perhaps the simplest method of stopping garden animals from eating a crop is a **physical barrier**. If you live in the country you may need to fence the entire home garden against rabbits and deer. A simple trick stops rabbits from burrowing under the fence. The rabbit mesh is attached with the bottom 30cm (12in) bent out to lie along the ground (there is no benefit from digging it in). The rabbits try to dig at the base of the fence but are stopped by the mesh on the ground. Fortunately they never figure out that they just need to back up a little. A deer fence needs to be at least 1.8m (5.9ft) high to stop them from jumping it.

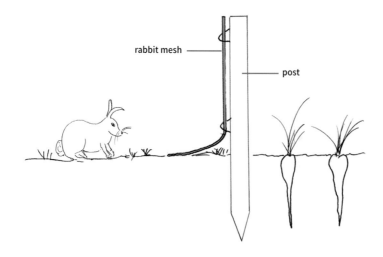

rabbit mesh

post

With birds the barrier needs to be in three dimensions, with the whole plant netted. I don't much like netting as it is expensive, usually plastic and runs the risk of trapping or entangling birds, so I only use it on the most vulnerable crops and only for the period of highest risk (although it is better to put it on too early than too late, before the birds get into the habit of coming to that plant). Making sure that netting is taut and well sealed at the ground reduces the risk of entanglement. If good care is taken a piece of netting can last for many years; patch small holes with thread and be careful never to put a net away damp.

Slug and snail barriers occupy a special place in gardening lore, with everything from copper pipe to crushed eggshell recommended to go around the base of vulnerable plants. Virtually none of these will really stop a determined gastropod – and once word gets out about the tender plant within they will all be very determined! The only thing that I have found really effective is a flared collar, which works not so much by deterring as by diverting our slimy friends. They climb up the collar, go round inside the lip a few times and get fed up. Like the rabbits with the rabbit fence they never work out that they need to go down and over the lip. I find them 100% effective – unless something like a leaf makes a bridge over the collar, in which case they are immediately rendered useless. They are only

Un-netted and netted currants

Slug and snail collar

needed for a short amount of time while a newly-planted plant hardens off, so I find that half a dozen of them are enough, moved around the garden as needed.

A home garden should already be well stocked with **predators** that eat many potential pests as the nectar and pollen in the flowering crops attract insects whose larvae are voracious aphid eaters, but there are many extra steps that you can take to increase your garden workforce. Often these are identical to the recommendations for increasing garden wildlife in general. A pond will attract slug-eating frogs, which also enjoy the cover provided by long vegetation. Leaves, prunings and clippings heaped up into a 'habitat pile' might attract a hedgehog looking for somewhere to hibernate. Rocks and pieces of wood (such as the boards used as paths over beds) encourage centipedes and predatory beetles like devil's coach horses, which find slug eggs a tasty treat. Insects like lacewings (whose adults look impossibly delicate but

Devil's coach horse

whose larvae are ferocious predators, from the point of view of an aphid) overwinter in hollow plant stems. You can encourage them by leaving dried-out stems over winter or by replicating them, using corrugated cardboard and a plastic bottle to build a 'lacewing hotel'.

Birds are ambiguous allies, eating both pests and the crops themselves. Many small birds live on a mix of seeds, fruit and insects. They need to balance the sugars from the fruit with the protein-rich insects, so allowing them to eat some of your fruit may encourage them to seek out insects and might be regarded as payment for services rendered. You can encourage birds by providing bird boxes, making up for the lack of ancient trees full of holes in our landscape. Where the birds themselves are a pest, I'm not suggesting that you encourage their predators, but our allotments do receive regular visits from a sparrowhawk, adding another level to our agri-ecological food chain!

While providing ideal homes for pest predators in your garden makes obvious sense, providing them for the pests themselves is a little more counter-intuitive! It may make more sense if you consider the difference between going around your garden at night with a torch hunting for feeding snails (as is traditionally recommended) and going to the place where they all hide out during the day. The majority of pests are nocturnal and if you know their preferred daytime haunts it is far easier to find them. If you have made ideal homes (for them) in convenient places (for you) it is easier still. Snails are particularly keen on gaps between

Bird box in the forest garden

Square bird nest from a bird box!

New Zealand flatworm, *Arthurdendyus triangulatus*

two vertical surfaces, such as a slate propped up against a wall. I have a number of these scattered around the garden. I collect snails every now and then and transfer them to my compost bins, where they chew up the composting material rather than my tender crops. Slugs are often found under stones or boards. New Zealand flatworms also hide under boards during the day, but seem to favour heavier objects with a flat surface against the ground, such as flat stones or large tubs. My mint tubs, made of decommissioned recycling boxes, double as flatworm traps. The same goes for any kind of pesticide, including natural ones.

It's also useful to practice good garden **hygiene** – but with the awareness that hygiene doesn't mean sterility. With personal cleanliness, scientists now promote the idea of 'targeted hygiene'. It is now recognised that the war on microbes promoted by the makers of anti-bacterial handwash and the like is actively bad for our health, preventing us from developing a well-trained immune system. On the other hand, a few pathogens are so harmful that they really do need to be avoided. Targeted hygiene focuses on these baddies while taking a more relaxed attitude to what old folk in my village used to call 'clean dirt'. The two sides of targeted hygiene actually work together, as a healthy population of harmless microbes takes up the space needed by the harmful ones and stops them from proliferating.

It is the same in the garden. The obsessive removal of weeds, pests and detritus removes the resilience from the system and leaves the gardener with an endless fight against invaders. However, there are some pests and diseases that you really do want to worry about and you should target your efforts on these.

Speaking of traps, I discourage any kind of lethal traps, such as beer traps or sticky traps. These are often indiscriminate, catching and killing as many beneficial animals as unwanted ones.

The first step in good garden hygiene is to be aware of what you are bringing in. New Zealand flatworms love the undersides of pots and have spread largely through the nursery trade. I now avoid the movement of soil and compost as much as I can, preferring to raise plants from seeds or by 'bare root' plants in winter. If buying a potted plant, tip out the rootball and examine it closely for passengers. Crates or trays of pots are probably worst of all. Similarly, reject any plants showing any sign of disease.

Within the garden, good hygiene is mostly about stopping pest and disease build-up in one spot. From insects to fungi, a lot of organisms cycle between the plant and soil, affecting the plant in the summer and then overwintering in the litter layer of soil. Annual gardeners rotate crops, planting them in a different spot every year, to avoid this, but this isn't an option with perennials. Instead we need to break the transmission cycle by removing affected material. This can be as simple as throwing a diseased fruit over your shoulder into another part of the garden, as I do when picking plums or apples. Alternatively, materials can be gathered up and composted. Apple scab overwinters on fallen leaves, so I gather them up and put them in the compost. The composting process may be enough to break down the pest or disease organism, but remembering to use the compost in a different part of the garden also helps. With cane fruit like raspberries, removing the dead stems at the end of their second year and using them in paths helps to prevent the transmission of disease from one year's canes to the next.

At the same time, it is worth remembering that a lot of disease agents, like the spores of rust and mildew, are ubiquitous, so there is nothing to be gained from treating diseased material like toxic waste. Here growing conditions are more important, with still air and wet promoting growth. Plant and prune to encourage air circulation and don't put vulnerable plants where fences and walls impede air movement.

Finally, remember that garden pests have another name: nature. In segregated vegetable and wildlife gardens we can pursue one aim in each, but in an integrated home garden it is more a question of finding a balance. We are cultivating a rich, productive system and tapping off just what we need from it. The rest can be given back.

Pruning
Maintenance pruning

Forest gardening draws on the way in which nature itself works for many things, from fertility to pest control. Therefore, when it comes to pruning, you might imagine that it would be best to leave aside such artificial practices and instead let trees and shrubs follow their own natural patterns of development. No-work gardening at its finest!

This was my own assumption when I started forest gardening, but years of experience have taught me, often the hard way, that the opposite is true. Pruning is a very effective intervention that pays back the effort many times in terms of the accessibility, quality and size of your crops. It is also a great way of learning more about your plants and how they develop. With a little practice, you can use your secateurs as a baton to conduct a symphony of growth, flowering and fruiting in your garden.

I use pruning to achieve four main things. The first is **access**. The secret to a forest garden is its complex, multi-layered structure. This can also be a hazard as you battle your way through branches and wet foliage to get to your food or get poked in the eye bending down to pick from the ground layer. The main principle here is to prevent plants from branching too low down so that you have a clear view and easy access to the plants growing beneath them. You can also make trees and shrubs more robust and so less likely to be dragged down by the weight of their own fruit.

The second is **yield**. Pruning is your main method for persuading trees to put more energy into fruiting and less into massive amounts of woody growth. It largely isn't a matter of cutting off growth after it has happened but instead of cutting back to the right buds so that you encourage the

kind of growth that will bear fruit. A tree that is investing less in vertical growth will have more to put into fruit and the lighter, sunnier canopy will generally allow better ripening of the fruit. In a forest garden, there is also a knock-on effect in the layers below. Many plants naturally produce more foliage than they strictly need for photosynthesis in order to shade out the competition and claim a greater share of the available water and nutrients. Creating a lighter crown gives you more light and productivity in your ground layer.

The third factor is **strength**. Modern varieties of fruit trees have been selected for both size and quantity of fruit, resulting in their branches having to support a far greater weight than their wild ancestors evolved for. There is no point in leaving the natural tree shape to support this unnatural weight. Pruning can be used to produce shorter, stockier, stronger trees.

Finally, there is **disease**. Pruning is an opportunity to cut out diseases like apple canker and to stop them from spreading. A more open, airy plant is also generally less prone to diseases such as mildew that can attack both plant and fruit. A pruning saw is a double-edged sword though. Carelessly-carried-out or mistimed pruning can spread infections or create opportunities for diseases to attack. I generally wipe my pruning tools with meths between trees to avoid transferring diseases.

There are three main principles to bear in mind when pruning. Much of the rest flows from these three. The first is that a bud will produce a shoot pointing in the direction that the bud itself is pointing. On a horizontal branch for instance, a bud on the top of the branch will produce vertical growth, buds on the side will grow out horizontally and ones on the bottom will grow down. The second principle is that growing buds suppress the growth of ones further down the shoot. The practical upshot of this is that you can cut a shoot back to a particular bud and this will be the one that will have the most growth. Leaving the end (or apical) bud will give strong extension of the shoot. The third principle is that plants prefer to fruit on some kinds of growth more than on others. The details vary from plant to plant and the biggest differences in pruning strategy are down to this.

With these principles in mind, let's have a look at some of the different species. I'll describe pruning methods here rather than in the species write-ups as there is overlap between some species.

Raspberries (*Rubus idaeus*) are often grown up wire frames that support the canes and prevent them from being dragged down by the weight of the fruit. The downside of this approach is that canes that wander too far from the frame are usually pruned off, preventing the natural spread of the plants. Over time this tends to lead to the loss of the whole stand to disease.

An alternative approach, which I have now adopted for all my raspberries, is to shorten the canes somewhat so that they are robust enough to be free standing. Allowing my plants to wander around the garden has allowed me to make some interesting observations. Raspberries are related to brambles (*R. fruticosus* agg.) and while they don't share the briar's trait of making roots at the end of their shoots in order to spread through a wood at

terrifying speed, they do share its ultimate dream – of getting up into a tree in order to co-opt its structure and reach the sun. The thorns of bramble and raspberry, like rose, both point backwards from the stem. They are as much about helping the plant to get a grip on the plants that it scrambles over as they are about protection – they are little grappling hooks designed to pull the shoots up into a tree.

Raspberry canes grow with a distinctive kink near the top. When they get near a tree, the purpose of this becomes clear: the canes first punch up vertically through the canopy then loop over, hopefully snaring a branch for support. In the open, I generally prune the canes just below this bend. It doesn't remove too much stem but it seems to be enough to keep them vertical. Under a suitable tree, I just let them get on with it. The pruning is carried out in winter when the plants are dormant. At the same time I remove the previous year's canes, take out any weak, spindly, damaged or diseased canes and thin out any areas that are too crowded.

The exception to this pruning strategy is for autumn-fruiting rasps, which have to be cut right down to the ground at the end of every season and

Raspberry canes with the distinctive kink at the top

which generally need some support once they are grown up.

Brambles or blackberries are like raspberries, except that their shoots are designed to arch over and root at the tip, which makes them something like a cross between creeping buttercup and barbed wire. However, it's possible to tame a bank of brambles with a little carefully timed pruning. In the middle of summer, when the shoots are growing at their fastest, cut the new stems in half, at the high point of their arc. If you've done the same the year before it is easy as you cut them just beyond where they pass the fruiting stems, so it can be done with a hedge trimmer if you have a lot.

The **currants** (*Ribes*) nicely illustrate the importance of knowing what kind of wood a plant fruits on. Blackcurrants (*R. nigrum*) fruit on new growth so they are pruned by cutting the oldest third of the stems down to ground level every winter, giving a constant supply of new shoots.

Redcurrants, whitecurrants, pinkcurrants (all *R. rubrum*) and gooseberries (*R. uva-crispa*) all fruit on old wood, so the pruning strategy for them is completely different. Winter pruning is limited to taking out especially old or diseased branches and any shoots that are crossing or growing up from the base. Then, in early summer, all side shoots can be cut back to two buds. Because the plants fruit on older growth this doesn't remove any fruit, but does change the plant's priorities from making new growth to producing fruit. Leading shoots can be shortened to create a robust, compact plant that will not be dragged down by the weight of the fruit on it.

Overall, the aim with redcurrants and gooseberries is to produce an open, 'goblet-shaped' plant that allows air circulation and resists mildew. They are usually grown on a 'leg' (the stem of the goblet), a short section of trunk free from any shoots. This is particularly useful when you have a ground layer crop that you want to harvest and also helps to discourage gooseberry sawfly, which emerge from the ground every year and have to get up into the plant in order to strip its leaves.

Apples and pears are pruned in the dormant season, between leaf fall and bud burst. The overall aim of pruning is to produce a robust, compact frame of sturdy branches and short fruiting 'spurs' that will support the unnaturally large fruits that we have selected the ancestral apple for. Spurs are side shoots that are kept short (just a few buds). As well as ensuring that the apples are borne close to the main stem this encourages the formation of fruiting buds.

The other main aim is to stop the tree putting all its resources into non-fruiting growth. The rule of thumb here is that vertical stems grow rapidly and produce little fruit while horizontal stems grow slowly and produce lots of fruit. As I said above, you don't want to have to achieve this by cutting off lots of the wrong kind of growth but rather by pruning back to buds that are going in the direction that you want them to in the first place.

Before you start, it's important to learn to distinguish fruiting buds (which will break to produce a spray of blossom, followed by fruit) from vegetative buds (which will produce a length of stem). In the photo opposite you can see some spurs pruned back to the large fruiting buds. The vegetative buds on the side of the stems are much smaller. Before you start pruning, check whether your tree is a spur-bearer, with the fruit buds closer to the base of the shoot, or a tip-bearer, with the fruit buds clustered at the end of the shoot. Spur-bearers are more common.

The first part of the pruning strategy is simple: reduce all vertical shoots to short spurs (or even cut them right back to the branch as they will always tend to want to make vertical growth). The only exception to this is for any stems that you want for vertical extension of the tree in the future, if any. These are generally reduced by about half to keep the tree compact and the stems robust. Also cut out any diseased wood at this stage.

The second part of the strategy is to create a leading shoot at the end of each branch that will

Fruit buds on spurs

Breaking bud

Stem grown from a downward-pointing bud

give horizontal growth. Pick a shoot (or two) near the end of the branch and prune it back, by about a third to a half, to a bud on the underside of the shoot. This will grow in the direction it is facing, giving a downward-facing shoot (the tendency of all shoots to curl upwards will translate it into a horizontal one). If all shoots on a branch are trimmed back hard then lots of buds will break, setting you up for a dense tree with lots of vertical growth in the future. The end shoots therefore achieve two aims: they give you a horizontal extension to your branch and they produce chemicals that suppress the growth of the buds further down the branch. The photos show the results with a shoot that has been pruned back to a downward facing bud at the start of the year, resulting in a horizontally-growing new shoot.

Finally, the side shoots behind the leading stem are your spurs. Here you can take your pick. Traditionally, spurs were cut back to a short stem of approximately five buds, ending again on a downward-facing bud. More recent thinking is to do no more cutting back than is needed to avoid congested and crossing spurs. If the fruit buds are clustered at the tips (a tip-bearer), this is definitely the option you need. The weight of the first year's fruit will pull the spur down to set the spur as horizontal, fruit-bearing wood.

Avoid pruning stone fruit like **plums or cherries** in wet weather or in winter as this makes them vulnerable to a disease called peach-leaf curl. A sunny day in July is ideal. Plums do not require a lot of pruning – mostly just cut out any diseased branches and any shoots that are growing vertically or towards the centre of the tree. Summer pruning may involve removing some shoots with developing fruit. Don't worry about this. Plums always set too much fruit anyway. In fact, you might want to thin out the developing fruit at this stage to avoid overcrowding, broken branches and exhausted trees.

An alternative to pruning on apples, pears, plums and cherries is **festooning** – tying down vigorous vertical branches to the horizontal. This is more work but gives a quick way to achieve a more horizontal structure without removing lots of material and making lots of cuts. I use it, in combination with pruning, on trees that haven't been pruned for a few years and need some drastic action.

Hazel trees also benefit from pruning but they are a different story from the fruit trees. Maintaining a tree with a single trunk and side branches would be difficult as the hazel is a determinedly shrubby tree. It is pruned rather like a giant blackcurrant.

Upright stems are 'stopped' at around two metres by cutting them back to an outward-facing bud or weak lateral shoot. This encourages the side shoots which bear the catkins and flowers. The aim is to have an open centre with no more than 10 main shoots. Up to a third of the oldest shoots can be cut out every year, leaving a 2-3cm (0.8-1.2in) stub for new branches to grow from.

Suckers and any inward-growing shoots should be removed.

In August hazels can be 'brutted'. The longer side shoots of this season's growth are bent, six to seven leaves along from the base, until they break halfway through. They are not broken off completely but are left hanging on the tree. This reduces woody growth and also lets more light into the tree, encouraging female flowers to form. Come winter, the broken parts are pruned off.

Formative pruning

If you are pruning a newly-planted fruit tree, you don't need to worry about encouraging fruiting as you shouldn't allow it to fruit in its first year. Instead, this is the time to think about the future structure of the tree. The aim is to produce a strong structure, with a clear length of trunk topped by

Brutted stems

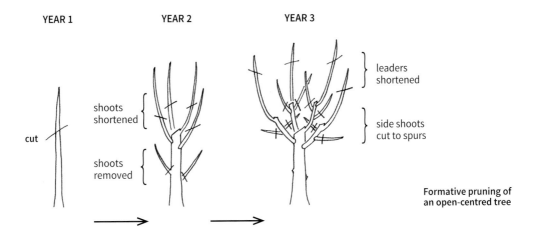

Formative pruning of an open-centred tree

shoots that are well-placed to grow on into main branches that are at a good angle and not too crowded. The trunk length can be anything from 75cm (30in) for a small 'bush' tree, to 1.2-1.5m (3.9-4.9ft) for a 'half standard' and 1.8-2.1m (5.9-6.9ft) for a 'standard'. In the forest garden it helps to have a good length of clear stem for access to the ground layer, so I like at least one metre. Above this point, fruit trees are usually pruned to have an **open** shape, with no central stem.

Formative pruning takes place over the first three years of a tree's life, after which the branches will be strong enough to support the fruit. If you buy a fruit tree from a nursery the formative pruning will often have been done for you, in which case you can skip some or all of these steps. Trees can also be bought either as 'maiden whips' with just one central stem, or as 'feathered maidens' with side shoots.

The first step is to get the tree up to the branching point. You will probably obtain the tree already tall enough, but if not, trim any side shoots to half their length to encourage vertical growth. The trimmed shoots should then be cut right back to the main stem at the end of the growing season.

In the first winter after the tree is tall enough (which is usually at the time of planting), the next step is to cut the leading (central) shoot to a bud at the desired branching height above the ground.

The following growing season the tree will produce side shoots growing out from the main stem and curving up. In the second winter, we want to select three or four of these to be main branches of the future tree. Look for shoots that are high on the stem, a good size, at a wide angle to the stem and arranged evenly around it. Trim off any that are low down, damaged, too small, close to vertical or crowded. The selected stems are cut to an outward-facing bud about one-third of the way up their length.

In the next growing season, the outward facing buds will grow long shoots. These 'leaders' are cut to an outwards-facing bud about two-thirds of the way up their length. Other buds will also grow to produce side shoots. These should be cut back to three to five buds to start forming spurs. The framework of the tree should now be strong enough to bear fruit and we transition into maintenance pruning.

It is also possible to prune a fruit tree to grow with a **central leader** (think of a Christmas tree shape). This is done, for example, with cider apple trees where the fruit is shaken from the tree rather than picked, and being able to reach it is less important. Unlike an open, multi-stemmed tree, the central leader is allowed to keep growing until the tree has six to eight good, well-spaced side branches above the desired height. Side shoots below this height or that are too crowded are trimmed to a few buds in their first year and removed in their second.

The branches that are retained are pruned to an outward-facing bud about two-thirds along their length (i.e. a third of the year's growth is removed) each year to leave a shorter, stronger framework for supporting the fruit.

Grafted trees

Occasionally, the rootstock of a grafted tree will put out its own top growth. If left unchecked this will rapidly outcompete the grafted variety, so you need to prune it off. See p.58 for using grafting to propagate your own varieties.

Cutting tools and technique

Secateurs and their larger cousins, loppers, come in two kinds: anvil and bypass. In the anvil arrangement the cutting blade presses down on a flat surface opposite. Anvil loppers are more robust than bypass ones but they damage stems more, so they should only be used for cutting up material that has already been cut from the plant. In the bypass arrangement the cutting blade slides past a thin, flat surface. The opposite surface serves to hold the stem as it is cut. This is the arrangement that you want for pruning.

The key to using bypass cutters is that the blunt side should be up the stem from the cutting side, that is to say on the part of the plant that is being removed, not the part that will remain. This is because the blunt side crushes the stem somewhat as it holds it. If this tissue damage is on the remaining stem then it is likely to die off at the end. If it is on the part that is cut off it does not matter. The blade causes hardly any tissue damage on its side, leaving a nice clean cut.

Secateurs and loppers are easily damaged by trying to cut stems that are too large. For these, a retractable pruning saw is an essential part of the forest gardener's toolkit. The thin blade of a pruning saw is ideal for getting into the crowded space of a tree's crown. Pull the branch gently away

Bypass secateurs

from the cut so that it stays open and doesn't trap the blade. Make a small cut on the opposite side from the main one at the point where the cut will emerge; this stops the bark from pulling off at the end of the cut.

With all cuts large and small it is best to avoid leaving a horizontal surface. This prevents water and fungal spores pooling on the cut. Pruning cuts are usually done just above the bud, sloping away from it.

BYPASS CUTTERS

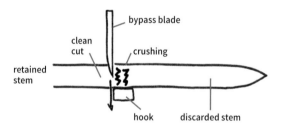

ANVIL CUTTERS

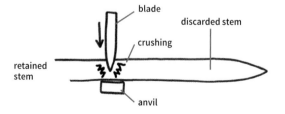

Supporting broad beans with stakes and string

Supporting your plants

In many ways, the ground layer of a home garden is quite like a traditional Victorian herbaceous border, which relied on a lot of supports to stop plants flopping over on top of each other. Nowadays plants for herbaceous borders have been bred to be shorter and more robust, but we mostly don't have this luxury in the forest garden. Some support for your plants is likely to be essential. There is no point in growing lots of lovely food if it is all laminated to the ground.

Plants rely on mutual support to a large degree, so if you support the floppiest plants you will often find that the rest stand up. Methods for doing so are lifted straight from the herbaceous border. You can stake individual plants with a strong central stem, make a 'cage' by putting several stakes around a group of plants and joining them together with string (which works well for

Dwarf hops climbing up an artichoke

73

Edge supports

multi-stemmed plants or groups of plants – broad beans are usually grown this way), or fix a piece or rigid mesh horizontally above the ground and allow plants to grow up through it. Alternatively, you can pair a robust plant with a climber (such as artichoke and hops) and use one to support the other. This does have to be done carefully to avoid the climber overwhelming the other plant.

To avoid plants flopping into the paths, you can plant the shortest plants at the path edges, gradually tapering up as you get towards the centre of the bed. This arrangement is also good for access. Alternatively, you can plant robust plants or place supports along the edge of a path, so that they

support everything within the bed and you walk down a canyon of plant growth.

Using the prunings

Forest gardens produce a lot of wood, occasionally in the form of large trunks or branches but more often as twigs and sticks. I have already discussed options for disposing of these or feeding them back to the garden, but another option is to use them as fuel for outdoor kettles and stoves, making the home garden more homely.

I have two pieces of equipment designed specifically to make use of small diameter wood in an efficient and clean-burning way. The first is a Kelly kettle, which consists of a small fire dish over which a chimney made from a water-filled metal jacket is placed. This keeps the flame and the water in contact via a highly conducting surface over a large area, and it is astonishing how quickly you can boil a large flask of water with a tiny handful of twigs.

Making a brew with a Kelly kettle

The other is a rocket stove. They vary greatly in design, but the essence of a rocket stove is a highly insulated burn chamber, with the outlet for the flame separate from the air intake. Long sticks are pushed slowly into the burn chamber or, in more sophisticated models, feed themselves in by gravity as they burn. The design means that the wood is burned completely and all the heat generated goes out of the top of the stove. Compare this with a camp fire, where a lot of the gases and soot particles generated by the fire go unburned, and most of the heat goes out of the sides. The result is a flame reminiscent of a gas cooker, perfect for putting a pan or a wok on top of.

Rocket stove

Stems peeled for candying

5

Eating from
the Forest Garden

I have found forest gardening to be at least as much of a culinary adventure as a horticultural one. The key to unlocking the potential of a plant is usually the discovery of the right way to cook it. Making the most of home garden produce means adapting in a number of ways. Most of the plants are unfamiliar, meaning that we need to get used to an array of new flavours, and a harvesting session can easily involve picking from dozens of different plants, so we have to learn cooking techniques that can cope with this diversity.

Home garden produce is highly seasonal. A few crops yield for a long season or even most of the year, but others can be available for no more than a week or two as they grow through a crucial, edible, stage in their development. If you want to live largely on the produce from your garden, you will find a huge contrast to the everything-available-all-the-time supermarket diet. For me this is largely a plus. The mild disappointment of not being able to get, say, wild garlic in August, is far outweighed by the pleasure of anticipation as new crops become available throughout the year and the reconnection with the seasons that it brings.

Trying a new food

Eating from a home garden means often trying new foods. There is more of an art to this than you might imagine. Our bodies err on the side of being suspicious of new substances – and for good reason. Our subjective experience of this caution is that the food tastes bad. If we force ourselves to eat a large plate of something we don't like, that bad experience becomes linked with the taste of that food and it will taste even worse next time we come across it.

The best way to accustom our taste buds to new flavours is exactly the same as the best way to get young children to like new foods: try a small amount of the food repeatedly. After a while your body will realise that the food is harmless and this will show itself as a change in your experience of the taste. Many of the foods I look forward to most keenly in the year didn't taste good to me at first. In an extreme example, Turkish rhubarb smelled so bad at first that I couldn't even bring myself to try eating a small piece. For months I did nothing more than rub my hand up a stem whenever I passed it and give it a sniff. Eventually I acclimatised and now I enjoy the flavour: more savoury and complex than that of garden rhubarb.

Another reason not to rush into trying a large plateful of a new food is that our reactions to food are very individual and you never know if you might react badly to a new plant. It is also just possible that you don't even have the right plant; even reputable nurseries make mistakes and labels can go astray in the best-managed botanical garden. Plants also vary and some can have higher levels of various substances than others (this is as true of commercial crops as it is of home garden ones, and new varieties have to be tested before being put on the market). Make sure you have positively identified any plant before eating, and only ever try chewing and spitting out a small piece in the first place. Don't try more than one new plant at the same time. Listen to what your body tells you, especially if the food is unpleasantly bitter or acrid. Bear in mind that many foods need proper preparation to make them palatable and read up on preparation guidelines before trying anything.

I don't want to make this sound more dangerous than it is. I have never poisoned myself despite trying literally hundreds of new foods in the course of my foraging and home gardening experiments. A little caution is sensible, that's all. The reward for this effort is food with more richness of flavour than you could find in the fanciest of restaurants, healthier than the dullest fad diet, and all for free.

A low-impact diet

I am often asked if I am self-sufficient with my forest garden. My answer is that I am not aiming to be. I am aiming to reduce my impact on nature in how I get my food, and self-sufficiency as an aim in itself does not necessarily help here. Foods that we buy vary hugely in their environmental impact. Some plants are simple to store and transport without special treatment and energy input – especially those which dry easily. Examples are beans and grains, some root vegetables and bulb alliums. Others can only be brought to market with refrigerated storage and transport, masses of packaging and air freight. Local is not always better; bananas' biodegradable, built-in packaging system and suitability for sea freight make them lower impact than supermarket strawberries. My priority is to replace these high-impact foods.

In *The Transition Handbook*, the permaculture author Rob Hopkins talks about 'the cake analogy' in local production. The 'cake' is the bulk items that make up most of our diet, while the 'icing' and the cherry on top are the luxuries like spices that can't be sourced locally. In the past the cake items would have been grown locally while the icing would have been imported. Now, if anything, the situation is reversed. Cake items are sourced from wherever in the world they can be produced most cheaply, while only luxuries – the local, organic, hand-crafted artisan whatever – can be made locally. International transport of food has its place, but its impact would be far lower if it was for the low-weight, high-value icing only. I try to apply the cake/icing perspective in cooking from the home garden. I buy very few vegetables, but don't fret about buying soy sauce from China to go on them.

Many people are now adopting a vegan or vegetarian diet to reduce their burden on nature. There is absolutely no reason why forest gardening and vegetarianism have to go together, but if you are following a plant-based diet then a forest garden is a great help, extending the variety available without relying on imported, heavily-processed foods.

Useful cooking techniques

Ways of cooking individual ingredients are covered in the plant chapters, but there are some general cooking techniques that are useful across the board. There are three main considerations in cooking home garden ingredients. The first is their diversity: cooking techniques that deal with a wide range of ingredients are essential. The second is the composition: expect more leaves, shoots and fruits than in the normal diet. The third is flavour: you can have too much of a good thing, and some ingredients need preparation to tone their taste down to be pleasant rather than overwhelming.

My favourite preparation method is **stir-frying**. It's a technique perfectly designed for using lots of small amounts of ingredients. The foods have to be sliced into thin strips for fast cooking – and many shoot vegetables come the right size straight from the garden.

The key to stir-frying is to prepare all the ingredients in advance of cooking. They are then added to a wok on high heat one by one, stirring to turn in each ingredient, working from those that need most cooking to those that need least. The exact order depends partly on taste, and the best order for you is a matter of experimentation. Noodles, if added, go in last; I find that spaghetti or linguine work just as well as Chinese noodles! At the end, liquid ingredients are added to cool the pan and add flavour, the lid is put over the wok and everything is simmered for a few minutes. Wine, light soy sauce and lemon juice give a Chinese flavour while stock and a few lime leaves add a Thai twist.

Leaves can be cooked like **spinach**. An onion and a clove of garlic are fried in a pan, then the washed leaves and a little water are added, the lid is put on and the leaves are steamed until soft. This is suitable for plants with large, soft leaves. It also drives off or breaks down strong-tasting components in leaves like wild garlic or nasturtiums, often leaving an improved flavour.

You can go a step beyond this recipe to make **leaf sauce**. This is suitable for a wider range of leaves and shoots. The ideal is a balance of milder leaves and shoots like kale, sea beet, spinach vine and sow thistle, and stronger-tasting herbs and shoots, such as lovage, sweet cicely, alexanders, mint and rosemary. Onion and garlic are fried again, then dry spices like cumin, coriander and white pepper can be added and fried briefly before adding the shoots and leaves, roughly cut to about one inch across the grain of the leaves. They are steamed for about 20 minutes until tender and then blended with a food processor. At this stage flavourings can be added, such as dark soy sauce, bouillon powder, yeast,

Preparing ingredients for stir-frying. Space can become an issue!

lemon juice, chutney – the only limit is your imagination. Don't, however, heat the sauce more than gently after this stage, as it has a tendency to bubble explosively, like a volcanic mud pool!

Leaf sauce is very versatile. It can be eaten on its own, with pasta or potatoes, as a base of a curry or layered into a lasagne. Leftovers usually get turned into soup the next day. It is very nutritious, and an old flatmate of mine used to find that it cured his migraines!

Soups and stews are also great for using a diversity of ingredients, especially the mix of roots, shoots, fruits, beans, mushrooms and tubers available in autumn and winter. Western cuisine has gotten out of the habit of using fruit in savoury dishes, but it was popular in medieval cookery and still is in Eastern cuisines. It is a habit well worth rediscovering.

There are two strategies for coping with strong flavours and acidity or bitterness in cooking. The first is to tone them down. One method of achieving this isn't really a cooking technique but begins in the garden before the food is even picked. **Blanching** involves excluding the light from a plant as it grows, resulting in an elongated, whitened shoot with a milder flavour. Shoot vegetables that are traditionally blanched include rhubarb, sea kale and udo, but you can try it on any herbaceous perennial. One way to blanch shoots is to pile light-excluding material, such as soil, woodchip or leaves, over the crown of the plant in winter. The shoots will be blanched only to the height of the pile, so it needs to be deep to be worth it. The other method is to use a blanching pot: a large pot that goes over the crown in early spring before the shoots grow. You can buy clay blanching pots specially for the purpose, but I find that an upturned bin, weighed down with a few bricks, does the job just as well. Be aware that blanching weakens the plant and makes it more vulnerable to slugs, so don't overdo it if you don't want to kill the plant.

A variation on blanching, known as **forcing**, is to bring the roots of biennials or herbaceous

Leaves for leaf sauce

Monk's rhubarb being blanched

Blanched vs unblanched stems of monk's rhubarb

perennials into a warm, dark place over winter, resulting in out-of-season blanched shoots like the chicons of chicory. Roots are dug in autumn and stored somewhere cool until they are ready to be forced, or in areas without frozen soil or thick snow

Lovage (blanched), udo, sweet cicely and hogweed shoots for tempura

they can simply be dug as needed. They are then trimmed of any remaining foliage, potted up like a pot plant and grown at room temperature with light excluded.

In the kitchen, acid or over-strong flavours can be tamed somewhat by coating ingredients with an **oily layer**. Cream on fruit is one example, or many shoots are nice simply fried in butter or olive oil, with just enough added water to stop them from sticking and burning. Some leaves absorb quite a lot of oil on frying and go crispy: possibly not very healthy but very nice. There seem to have been few things that coastal cultures in the Pacific North West wouldn't coat in fish oil before eating – which may have been an acquired taste but presumably achieved the same purpose. In the Southern US states the equivalent is frying greens in pork fat. A thicker coating can be achieved by the many forms of **battering**, from fritters to pakora to tempura, or leaves can be wilted and then fried in **egg**.

The opposite strategy is to harmonise strong flavours with other, equally strong ones. Thai **yam** salads make assertive tastes like raw onion more palatable by coating them in a dressing of oil, soy sauce, sugar, chilli and vinegar or lime juice. One of my favourite ways of preparing strong-tasting shoots is to prepare them as a Korean **namul**. They are steamed, then dressed in a mixture of light soy sauce, sesame oil and lemon juice. **Curry** is perhaps the ultimate example of this approach.

Tempura

Vegetables prepared for fermenting

I also include **lacto-fermentation** as a cooking rather than as a preservation technique, as I do it more for the taste than to increase shelf-life. Although the food is not heated, friendly *Lactobacillus* bacteria 'cook' it by digesting molecules that we find hard to break down, to produce dishes like sauerkraut and kimchi.

Lactobacillus thrive in salty, oxygen-free (or anaerobic) conditions. Ingredients for fermenting are first shredded, then kneaded with salt and packed tightly into a crock or jar. The salt draws moisture out of the vegetables to create a brine that should completely cover them, bathing the ingredients in salt and excluding oxygen.

The ingredients must be fresh; this is not a way to perk up wilted leaves. Cabbage is the classic ingredient for lacto-fermentation and I usually put a portion of cabbage in mine, but any leafy vegetable will do, as will other small or sliced ingredients such as radish pods and roots. The mixture should end up not too wet and not too dry. Leaves like kale are too dry on their own; fruit such as apples will add moisture but too much will overwhelm the mix and dilute the brine. Spices like ginger and chilli can also be added for a kimchi-esque ferment. Enough salt is then added to make the mix too salty for other bacteria, but not too salty for

A snugly-fitting lid in a fermenting crock

Fermenting nicely

Lactobacillus: 15g of salt to 1kg of vegetables (0.5oz of salt to 2.2lb of vegetables).

Kneading the shredded, salted vegetables is an important step. It softens the ingredients and allows them to pack tightly with no air gaps, and it starts to draw juice out of them to form the brine. The vegetables are then packed into a container, squashing each layer down with a fist to ensure that all air is squeezed out. Some way of holding the shreds below the surface of the brine is needed: a plate, slightly smaller than the container, with a weight on top of it does the trick nicely. A tea towel over the top keeps dust and flies out. If the container is sealed it is necessary to 'burp' it frequently as the bacteria release gases and pressure builds up.

The mix needs to ferment for around a week at normal kitchen temperatures. You have to keep an eye and a nose on it throughout this time. Good ferment will fizz somewhat and start to smell tangy. Bad ferment smells, well, bad. Any scum that forms on the surface should be skimmed off. The longer it ferments the sourer it will get, which is a matter of taste. When it is ready, jar it up (pressing in tightly again) and keep in the fridge.

Preserving your harvest

With a home garden's year-round production there is less need to preserve produce, but some preserves are worth making for their own sake.

Most fruits are suitable for making **jam**. If you want it to set, fruit high in pectin must be included in the mix, such as currants, gooseberries, apples or lemon juice. My overall jam recipe is to add one part sugar to two parts fruit; cook, stirring occasionally and slowly turning down the heat until the mix acquires a jammy texture and starts to spit; then jar up. Jam with this amount of sugar will keep indefinitely in a closed jar, but needs to be kept in the fridge after opening.

The savoury equivalent of jam is **chutney**. You'll find a chutney recipe that can be adapted for other ingredients in the section on rhubarb.

Robust, hollow stems can be preserved (and transformed) by **candying**. The name put me off this process for many years, but if anything it uses less sugar and energy than jam and really is the best way to make a few products edible, especially the thick flower stems of angelica. It also works for related species such as sweet cicely, lovage and alexanders, and even the flower stems of rhubarb. Stems are boiled and then steeped repeatedly in sugar solution, until the sugar has saturated them completely. The sugar draws water out of the cells and the repeated boiling drives it off and concentrates the solution. The hollow nature of most flower stems aids the solution in getting to all parts of the flesh.

LOW ENERGY STERILISATION FOR JARS

Jars can be sterilised for preserves using boiling water.

Step 1. Pour boiling water into the bottom of a jar (about a sixth the volume of the jar).

Step 2. Shake the jar, then release the pressure (wear oven gloves or use a thick towel for this).

Step 3. Pour the water into a flask.

Step 4. Refill the jar, shake, and pour the water into the next jar.

Step 5. Refill the jar, shake and set aside.

This process is repeated for the rest of the jars, with water from the previous jar counting as Step 1. The water in the flask can be reheated if more is needed or used to clean up at the end of the jam-making. Empty the jars, shaking out to dry, shortly before filling.

This method heats the jar slowly to avoid sudden temperature changes that could crack it, and leaves it hot enough to sterilise the jar. It can be carried out while waiting for the preserve to cook. *Jars must be hot before hot preserve is poured into them.*

Blanching stems

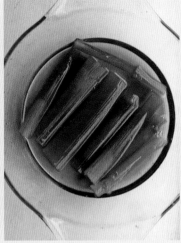
Stems in syrup after boiling

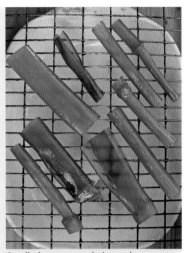
Candied stems on a drying rack

CANDYING STEMS

Step 1. Peel the stems and cut to a useful length for storage (p.76 photo).

Step 2. Carrot family stems can be blanched before candying in a pan of water with a teaspoon of baking soda in order to help preserve their colour. This step should be missed out for rhubarb.

Step 3. Bring half a litre of water and 400g (14oz) of granulated sugar per kilo of stems to a boil. Add the stems and boil for a couple of minutes, then pour everything into a dish.

Step 4. Leave the syrup to cool, then cover the dish and leave overnight.

Step 5. The next day, pour the syrup back into the pan. Bring it to the boil, add the stems and boil everything for a couple more minutes, then return to the bowl. Cover after cooling and leave overnight again. Repeat this process one or two more times.

Step 6. After the final boiling, remove the stems from the syrup, roll them into sugar and place them on a drying rack.

Step 7. Store in a cool dry place, in a paper bag or a jar with desiccant.

N.B. The remaining syrup tastes great and can be kept for use too.

Candied stems

Drying removes the need to add sugar or acid. Most fruits can be sliced thinly then put in a dehydrator until dry. Dehydrators come in two main varieties: vertical ones blow warm air up through a stack of trays, while horizontal ones blow air across them. The main difference between them is that all the trays in a horizontal dehydrator get the same drying conditions, while the bottom tray in the vertical type dries faster than the top one. If you want to do a lot of dehydrating I would suggest getting at least a four-tray horizontal model. Some products, such as apple rings, herbs and shiitake mushrooms dry without a dehydrator and only need to be hung up or laid out in a well-aired place.

A dehydrator can also be used to make **fruit leather**. Fruit is pulped, then spread on sheets of baking paper in a layer no more than 5mm (0.2in) thick and placed in the dehydrator. Dehydrating takes about 12 hours and can be started overnight. I have never met anyone who didn't like fruit leather, and it is a very healthy option as nothing is added to or taken away from the fruit apart from water. A piece of fruit leather slipped into a pocket makes ideal trail food.

Firm fruits such as apples may need to be cooked for a couple of minutes before liquidising, but most can be used raw. Very liquid fruits can be mixed with drier ones or with oatmeal and sour ones mixed with sweeter. You can also add spices: apple and cinnamon and plum and ginger are two of my favourite combinations. You can even be visually creative, with bands or shapes of different-coloured pulp. Fancy fruit leather gives lots of scope for creativity and makes a great present. Once dried and peeled off the baking paper (which can be re-used several times), fruit leather keeps indefinitely stored in a sealed container.

Finally, **fermentation** with either wild or added yeast both transforms and preserves fruits. Fermented drinks don't have to be especially alcoholic. My favourite use for it is to fizz up drinks like elderflower champagne and sparkling apple juice, often using the fruit or flower's own natural yeasts. The keys to fermentation are: an acid environment, either provided by the juice itself or with an additive like lemon juice; sugar, either from the fruit or added; and yeast. Many fruits and flowers naturally have yeast on them, attracted to the sugar in the nectar or the fruit, and for mild ferments this wild fermentation can usually be relied on. I once had a bottle of apple juice ferment naturally and kept the culture going for several years simply by topping up with more juice. This kind of fermentation knocks out some sugar and adds a sparkle, turning something okay into something spectacular. For stronger ferments, cultured yeasts are usually added, to deal with the stronger concentrations of

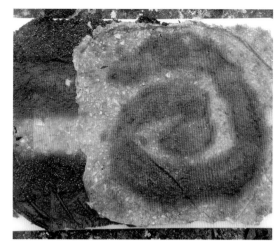
Patterned fruit leather

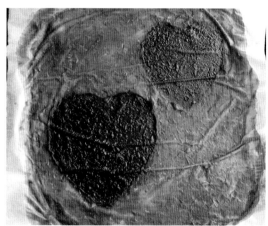
Fruit leather valentines

alcohol. The eventual potency of a brew depends both on the amount of sugar present and the strength at which the yeast poisons itself and stops working. For cider, baker's yeast will do; for wines or nettle beer, wine and ale yeasts that can tolerate higher strengths are needed.

Whatever you are making, don't look down on the lees, the layer of yeast and fruit cells left at the bottom once the main product is racked off. I keep a bottle of it next to the cooker as a 'secret ingredient' for soups, stews and curries. I use a small amount at a time to add 'bass notes' rather than as a main flavour, but it keeps forever since the liquid it is in oxidises to vinegar.

GENERAL FRUIT WINE RECIPE

Equipment:

Fermenting bin: made of food-grade plastic and with a tight-fitting lid with a small ventilation hole in it. Alternatively, use any food-grade container, with the top covered using a securely-attached plastic bag to allow expansion without letting flies in.

Cotton wool is used to fill the neck of a bottle or hole in a lid where you want to allow gases to escape without letting fruit flies in for a short while.

Demijohns (also known as carboys) are one-gallon (4.5l) glass vessels used as standard for fermenting and maturing wine. They can be made of clear glass, which makes it much easier to see what you are doing when siphoning the wine, or dark glass which protects red wines from being bleached by light.

A bag made from fine cloth, such as muslin, is used to filter the pulp out of the liquid when it is transferred from the bin to the demijohn. A funnel is also helpful at this stage.

An airlock goes in the top of the demijohn while the wine is fermenting. It allows CO_2 to bubble out while preventing oxygen or flies from coming in. Oxygen will oxidise the alcohol to vinegar.

Bungs, made from rubber or cork, are needed to seal the demijohn while the wine matures.

A rubber or silicone tube is used to siphon wine between containers without getting lots of oxygen in it.

Modern screw-top wine bottles can be re-used and work very well. If you're a traditionalist you'll need to buy corks and a corker.

INGREDIENTS:

For all wines

2-4kg (4.4-8.8lb) fruit, depending on intensity of flavour – less for strong-tasting fruit like elderberries and more for milder fruit like apples. Substitute up to 400g (14oz) raisins for fruit to add body and smoothness if desired.

4.5l water

1.1kg (2.4lb) sugar

Wine yeast. Different yeasts give the wine a different character. In general, choose red wine yeasts for red fruit and white for light-coloured fruit.

Yeast nutrient. Contains certain essential nutrients for the yeast. Add according to the manufacturer's instructions.

Steriliser: Campden tablets or 10% potassium metabisulphate solution (PMBS). These kill the wild yeasts from the fruit and the air. It is possible to leave this out, allowing wild yeasts to carry out the first stages of fermentation while the more alcohol-tolerant wine yeast takes over for the later stages, but it is riskier.

For some wines

Pectolase. Pectin helps jam to set, but it can make wine cloudy. This enzyme breaks it down. Use for high-pectin fruits such as plums, apples, currants, rhubarb and their relatives.

Citric acid or juice of two lemons. Yeast works best at a pH of 3-4. Most fruits are naturally quite acidic, but ones which are not, such as cherry, apple, pear, elderberry and rowan, can be acidified.

Precipitated chalk. Some fruit, like rhubarb or raspberry, can have too much acid. The acidity can be reduced with small amounts of precipitated chalk.

Tannin tablets or extra-strong tea (not including the tea leaves). Most fruits have tannin in them. It is the substance that makes your mouth pucker when you drink over-brewed tea or eat fruit like damsons or sloes. Tannin levels usually decline as fruit ripens. Some of it improves a wine, but not too much! It can be added to fruit that is low in it, like rhubarb, strawberries or whitecurrants.

Demijohn, tube and airlock

METHOD:

Day 1 – steeping

This step draws the flavour and sugars out of the fruit and into the liquid. An alternative to steeping is to juice the fruit; the juice can be fermented immediately, diluting if required for taste and adding pectolase and acid as necessary.

Wash the fruit and chop it into pieces if it is larger than berry size.

Place the fruit (and raisins, if using) in a clean fermenting bin.

Boil 2l of water and pour it over the fruit. Stir or mash lightly, cover and leave to cool to room temperature.

Add 1 crushed Campden tablet or 1tsp of PMBS.

Add 1tsp of pectolase to high-pectin fruits.

Cover and allow to steep for 24 hours. The result is called the 'must'.

Day 2 – beginning fermentation

Yeast is generally bought dried and should be 'reactivated' before adding to the must to get a fast start and outcompete wild yeasts. The must is then fermented in a fermenting bin until the process slows sufficiently to be transferred to a demijohn.

Boil 200ml of water and cool to lukewarm in a sterilised bottle.

Add 1 sachet or 1tsp of wine yeast and 1tsp of sugar and swirl gently. Stop the bottle with cotton wool and allow the yeast to rehydrate for 20 minutes.

Boil 2l of water and dissolve 1.1kg of sugar in it. Stir into the must.

Add 1tsp of citric acid or the juice of two lemons to non-acidic fruits. Alternatively, test the must with pH paper and add acid/juice to obtain a pH of 3 or

4. If the pH is lower than this, add precipitated chalk to bring it into the optimum range.

Add 1 tannin tablet or 1tbsp of extra-strong tea to low-tannin fruits.

Allow the must to cool to room temperature.

Stir in yeast nutrient.

Stir in the rehydrated yeast. Put the lid on the fermenting bin, with a little cotton wool in the hole in the lid. Leave the bin in a warm place for about 12 hours.

Open the bin and stir again. By this time the must should be showing some signs of fermentation. It should be bubbling and the pulp will be pushed to the top of the liquid. Leave for another 12 hours.

Next 4 – 14 days – main fermentation

Stir twice a day and keep an eye on how the fermentation is going, replacing the lid in between times. Once most of the sugar is used up, the must will stop bubbling and the fruit will sink to the bottom. It is now ready to be moved to a demijohn.

Finishing fermentation

Strain the must through a fine cloth bag into a sterilised demijohn, using a funnel.

Top up to the 'shoulder' of the demijohn with cooled, boiled water.

Add a bung with an airlock.

Pour a little water or PMBS solution into the airlock.

Leave in a dark place (or use a demijohn with dark glass), especially for red wines.

This is the fun part, as CO_2 produced by the yeast bloops away through the airlock. Fermenting demijohns make engrossing pets. Over a period of weeks or months, fermentation will stop, the bubbles through the airlock will cease and the fruit and yeast cells suspended in the wine will settle out, forming a layer called the 'lees' at the bottom. The wine is now ready for racking and maturing.

Racking and maturing

Once fermentation has stopped and the wine has cleared, 'rack' it by siphoning the wine into a new, sterilised demijohn, leaving the lees behind in the old one. Replace the airlock with a solid bung. You may need to rack the wine again if you get some lees with the wine, but don't do it too often as each racking allows oxygen into the wine. Allow the wine to settle thoroughly again and repeat the process. You can drink the wine as soon as it is clear or leave it to mature, in a cool dark place, for as many months or years as you can wait.

Bottling

When you are ready, siphon the wine into sterilised bottles. Sláinte!

PART 2

A DIRECTORY OF OF FOREST GARDEN PLANTS

6

Introduction to the Crops

The following chapters describe some of the many edible plants that can be grown in a home garden. It is by no means a complete list. I have tried to focus on those that, in my experience, are really and practically worth growing – although this is very much an individual judgement. I have also stuck largely to plants that I have some personal experience of. I have broken this rule mainly for two reasons: I have tried to mention equivalents in other parts of the world of northwest European plants that I grow in my garden; some others that I haven't tried just sounded too good to leave out. I have tried to make it clear where I am speaking from personal experience and where I am reporting the experience of others.

Sources of knowledge and inspiration

The sum of available knowledge about edible plants is increasing all the time. The species in a home garden are drawn from a variety of sources. Some have been cultivated crops for millennia. Some of those are familiar; others have fallen out of use in our increasingly mechanised and marketised food system but deserve to be rediscovered for home garden use. Other crops have been better known to foragers than to gardeners, and here we have to draw on the foraging traditions of the world for inspiration.

I have used a few sources too regularly to introduce them every time they are mentioned. One is Stephen Barstow, a Brit who has spent much of his life living in Norway. He says that on moving to Norway he discovered that Norwegians 'didn't believe in vegetables', and the only way to survive as a vegetarian was to start growing his own. Since then he has tried out seemingly just about every plant ever mentioned in the ethnobotanical literature (over 8,000 of them), with a particular weakness for the genus *Allium*. He made full use of his travels around the world as an ocean wave climate researcher to check out local plant use traditions, and distilled all this knowledge into the book *Around the World in 80 Plants*. Retiring as a wave scientist has only given him more time to work as a perennial vegetable researcher and he continues to publish fascinating snippets and monographs on the blog *Edimentals*. He also somehow finds the time to lead the Norwegian seed savers' network, KVANN.

The great majority of the world's knowledge about useful plants has been discovered and preserved by farmers, foragers and farmer-foragers who passed on their learning by word of mouth. Much of this information has been lost or come close to being lost due to colonisation and simple neglect. The work of ethnobotanists in recording these endangered traditions and making them available to the wider world has been invaluable. In the Pacific North West of North America, an area with many climatic similarities to North West Europe, I have drawn regularly on Nancy J. Turner's two volumes: *Food Plants of Coastal First Peoples* and *Food Plants of Interior First Peoples*. At the other end of the Americas, there is a smaller area of cool temperate climate in Patagonia, inhabited by the Mapuche people. The many papers by the Argentine ethnobotanist Ana Ladio and her colleagues are worth reading for a wealth of information on this little-known area.

The researcher Hu Shiuying grew up on a farm in the precarious flatlands of North China, where she says she acquired "first hand knowledge of many famine foods"[1] along with experience of the region's agriculture and horticulture. A scholarship allowed her to study botany and sociology at college, followed by a Masters degree in which she wrote a thesis on 'The Chinese Esculent Plants Used for the Conservation of Health'.[2] Her sources for this work ranged from Taoist monks and nuns to the shopkeepers and homemakers of Guangzhou. After the outbreak of the Sino-Japanese War she travelled as a refugee to Sichuan and became a lecturer in botany at the local university. She spent the war years teaching, botanising in this hotspot of plant diversity and staying in the villages of Sino-Tibetan ethnic groups. In 1946 she moved to the US and studied at Harvard, where she worked in the herbarium of the Arnold Arboretum and founded and directed the *Flora of China* project. In 1968 she became Senior Lecturer in Biology at the Chinese University of Hong Kong. Over the course of her 102-year life she produced over 160 academic treatises, collected over 30,000 specimens, and published the 800-page *Food Plants of China*.

Complementing the work of academic researchers, there is an ever-growing literature on foraging by active foraging enthusiasts. The one I quote most often is Samuel Thayer, the author of *Forager's Harvest* (2006), *Nature's Garden* (2010) and *Incredible Wild Edibles* (2017). There are two reasons for this. The first is that all Thayer's books are deeply grounded in practical experience, based on an obsession with wild foods that started in childhood. In a field where there is still too much copying of information from book to book without

any testing this is invaluable. The second reason is that he writes beautifully, and his books would be worth reading even if you had no interest in eating wild plants. His books cover North America, and not all of the species are from cool temperate areas, but a good proportion are.

One person who straddled the fields of academic researcher and enthusiastic forager was Gordon Hillman, a pioneering archaeo-botanist who studied the origins of agriculture in Southwest Asia. In his retirement he turned his attention to understanding the diet of Britain's ancient hunter-gatherers. Learning anything about such a long-vanished tradition required a lot of attempted recreations, applying techniques learned from archaeology and modern hunter-gatherers to Britain's native plants. This approach carries a degree of risk, and Hillman may possibly hold some sort of unofficial record for the number of times he poisoned himself in the name of science! His findings were published in the book *Wild Food*, co-authored with the bushcraft expert Ray Mears, and on the blog *Wild Food Plants of Britain*,[3] which continues to be updated by his colleagues since his death in 2018.

Anyone with any interest in useful plants will find themselves making regular use of the online database Plants For A Future (PFAF), largely compiled by Ken Fern. It aims to be a comprehensive record of published records about edible, medical and otherwise useful plants rather than a fact-checking service (although Fern's personal comments are often very helpful), so some scepticism and digging into the references is sometimes required. Due to the amount of copying from one source to another, what look like multiple independent records may in fact turn out to have just one original, sometimes obscure, source. One of the most useful features of the database is that it includes cultivation and propagation details alongside the information on uses.

Martin Crawford is the founder of the Agroforestry Research Trust, comprising a large demonstration forest garden next to the Schumacher College in Devon, a mail-order nursery of useful plants, and a website and series of books and pamphlets providing information on temperate forest gardening. Crawford is the author of *Creating a Forest Garden*.

Sometimes it is a concept or a tradition that crops up regularly instead of an author. This is most often the case in East Asia, where we find an almost unique combination of strong foraging customs and continuous, literate, uncolonised cultures that have preserved a living record of ingredients and cooking techniques. As a result, information is less often channelled through a single author but more distributed. With the spread of English in these countries and the growing comprehensibility of automatic translations, it is increasingly possible to access the information found in blogs, news articles and websites.

In Japan, we can look to the tradition of collecting *sansai*, or 'mountain herbs'. Japan's unusual geography gives it large wilderness areas in close proximity to large human populations, and people have always complemented agriculture with foraging from the forest, especially in the 'hungry gap' of spring. Sansai blur the boundary between food and medicine, with a parallel to the European idea of 'spring tonics'; they are not just calories but pick-me-ups carrying the vital force of mountain and forest to restore spirits run low by winter. In modern terms we might talk of essential phytonutrients run down by a winter subsisting on bland staples. In Korea, a very similar idea is expressed in the terms *san-namul* (mountain vegetable), *deul-namul* (field vegetable) and *bom-namul* (spring greens).

In China, this idea is taken one step further with *bupin*. Hu Shiuying writes, "Bupins are regular Chinese herbal medicines cooked in food and taken to cope with the stress and strain of life with environmental changes, and to replenish a person's natural immunity. Preparations made from these plant products are taken neither as food to assuage hunger, nor as medicine for curing specific ailments.

They are often taken seasonally and occasionally for keeping the body in tone and for the prevention of diseases."[4] At one end of the spectrum, bupins shade into the elixirs taken by Chinese emperors and others in a vain quest for immortality. At the other, they represent hard-won folk knowledge about the properties of plants, with many having been first tried in desperation in times of famine. I have taken traditional use as a bupin as one piece of evidence for edibility for a plant, but it shouldn't be taken alone to mean that a species is safe to eat in volume.

Understanding plant names and families

Before we get to the individual crop species in the next chapter, it is worth stepping back and looking at the larger groups that plants fall into. The nature of evolution means that species can be arranged into a vast Tree of Life, with the leaves, representing individual species, grouped together onto twigs, the twigs grouping onto branches and so on. It helps to get to know not just the individual species but these higher groups, such as the genus and family, as related species share common characteristics, making them much easier to remember and often giving a guide to their use.

The plants in the following chapters are named using their twofold botanical name, such as *Allium ursinum*. The first name is the plant's genus. A genus may contain only one plant or a number of closely related plants. The second name narrows it down to the individual species but is never used on its own, as many plants will share the same final name but no two will share the same two-name combination.

Scientific names are useful because, in theory at least, each species has only one scientific name and one scientific name refers to only one species. By contrast, a species may have lots of common names: *Allium ursinum* is known as wild garlic,

93

ramsons, ramps and bear garlic to name but a few. Common names may also refer to more than one species: 'ramps' is used for both the European *Allium ursinum* and the North American *Allium tricoccum*. Species sharing a common name may not even be that closely related. 'Sorrel' can refer to a range of weedy grassland plants in the genus *Rumex*, clover-leaved *Oxalis* species, a mountain plant called *Oxyria digyna* and even a tree called *Oxydendrum arboreum*.

In practice, however, botanical names are not as stable and straightforward as we would like them to be. Every time that botanists change their collective mind about which species belong together in a genus, a number of species end up being renamed. This has been happening a lot recently as genetic analysis gives insights into the relationships between species not available to earlier taxonomists. As a result, the name of a plant in older books may no longer be the one that it now goes by. Some have been landed with a change from a pretty, memorable name to something altogether clunkier, as in the move from *Aster scaber* to *Doellingeria scabra*. Fortunately, lists of synonyms are available at the likes of theplantlist.org, which is maintained by Kew Gardens.

It's also very useful to get to know the level of grouping above the genus: the family. The plants within a family are less closely related than those in a genus, but still closely enough that knowing a plant's family will give you a feel for its characteristics. These are some of the common families that forest garden plants fall into.

The celery and aralia families

The Apiaceae and Araliaceae are two quite closely related families; botanists may yet decide to amalgamate them. The flowers of the Apiaceae are usually white (sometimes pinkish) and consist of many tiny flowers held together by a branching

Carrot, a member of the Apiaceae

flower head that often holds all the florets in one plane, like a big, flat, white plate. They used to be known as the umbellifers from the name for this kind of flower head: an umbel. It is a large family, including everything from carrot to giant hogweed. The Araliaceae also have a branching flower head, but here the effect is more often like a starburst firework, with little spherical clusters of flowers around the final branching point (some Apiaceae also adopt this arrangement). They are common in North America and Asia, but strangely rare in Europe. Ivy is one native Scottish example.

Both families are versatile chemists, producing a dazzling array of tastes, poisons and medical effects. Carrots, parsnips and celery are in the Apiaceae, along with herbs like angelica, chervil, dill, fennel, lovage and parsley, and spices anise, asafoetida, caraway, chervil, coriander and cumin. And that's just the well-known ones! You'll find many more in these pages.

On the other hand, that same alchemical genius can also make the Apiaceae dangerous. It also includes some of the most poisonous plants around, including water hemlock, poison hemlock (the plant that did for Socrates), water dropwort and fool's parsley (a name that tells its own story). Even well-known food plants in the Apiaceae can bear keeping an eye on. Parsnips, like hogweed, produce a chemical that can cause dermatitis when it gets on skin that is exposed to strong sunlight. Luckily, having poisonous parts doesn't stop a plant being edible if treated correctly. Araliaceae put their pharmaceutical skills to better use: ginseng and Siberian ginseng are both members.

Most Apiaceae are annuals, biennials or herbaceous perennials. Araliaceae tend to be larger in size and include trees both large (*Kalopanax*) and small (*Aralia elata* and *spinosa*). Both genera are loved by pollinating insects.

The rose family

As you might expect of a family named after the rose, two characteristics of the Rosaceae are showy flowers and sweet scents. From the point of view of the home gardener, perhaps the more important quality is that many of them have fruits that are just as appealing. Apples, pears, quinces, apricots, plums, cherries, peaches, brambles, raspberries, loquats and strawberries are all members. So too are almonds, which have been bred for lower levels of cyanide in their seeds; a less fortunate trait of the Rosaceae. A small number of them also have edible shoots, in which the characteristic 'rosy' taste is reduced. Rosaceae range in size from small herbs to small trees. Leaves are usually simple or pinnate (leaflets arranged along a central stem). They are prickly customers, with thorns a common feature, but when life's a bed of Rosaceae, you can't complain about a few thorns.

The amaranth family

The amaranth family is a large one, and recently grew even larger by absorbing the Chenopodiaceae or goosefoots. Together they include a number of important leaf crops, including the amaranths, spinach, and all the derivatives of sea beet: chard, beetroot and sugar beet. Less well-known members to be found in this book include Caucasian spinach (*Hablitzia tamnoides*)

Wild rose flower

Quinoa growing in Peru at 3,800m

and saltbush (*Atriplex halimus*). Amaranthaceae are often annuals and biennials or short-lived perennials, sometimes small shrubs. They tend to diamond- or arrow-shaped leaves and many have a silvery bloom on the leaves which protects them from desiccation, adapting them for deserts and seashores. They are evidently nutritious: you can choose between fat-hen (*Chenopodium album*) and fat-husband (a Yoruba name for *Amaranthus*). Many are prolific seeders, which is good in seed crops like amaranth and quinoa, less good in the many weeds in the family.

Aster ageratoides

The aster family

The aster family is a cheery one, loved by gardeners and pollinating insects alike. Their sunburst flowers give them both their older name of Compositae (the flower is really a composite of many tiny flowers, with 'disc flowers' on a central disc and

Sunflower

'ray flowers' radiating out at the edge, looking like the petals of a single flower) and their newer one of Asteraceae ('aster' being Greek for 'star'). While the typical flower form is that of the daisy or sunflower, some have more thistle-like heads. Many Asteraceae are wildflowers and many are ornamental garden flowers, but what of their edibility?

A number of them have edible flowers, which tend to share the taste of one of their number: the artichoke. Juicy edible stems are common too, hidden under a bitter skin. Many also put down stores in enlarged roots or tubers, but unfortunately they have a habit of doing this in the form of inulin, the rather indigestible sugar that makes Jerusalem artichokes notorious for inducing farts. A few have edible leaves, which share a characteristic 'astery' taste. Some Asteraceae, like lettuces, dandelions, chicory, sow thistles and goatsbeards, do their best to protect themselves with a bitter latex that oozes out of leaves and stems when they are cut.

The Asteraceae is the largest family of all amongst plants. Members may be trees, shrubs or the fantastic cushion forms of Tasmania and New Zealand, but in the cool temperate north they tend to be annuals and small herbaceous plants. One cute habit that some of the family have is that of their flowers opening and closing with the day and tracking the sun's course round the sky. The sunflower is known as turn-to-the-sun in various Latin languages and the daisy draws its name from the Old English for 'day's eye'.

The heath family

The Ericaceae are better known as the 'heaths', and that is where they tend to be found. They are a shrubby family, adapted to acid soil conditions. Their leaves tend to be simple, small and waxy; an adaptation to regular water stress and life in the sun. Their main interest to forest gardeners is their berries: blueberry, huckleberry, cowberry, cranberry and salal are all members.

The bean family

After the asters and orchids, the Fabaceae (also known as the Leguminaceae, legumes or pea/bean family) contains the most species of plants. They can be anything from mighty trees to small climbers. The trees tend to be in the tropics; Scotland has no native leguminous trees, although we do have the imported laburnum and whole hillsides covered in broom and furze or whins (better known elsewhere as gorse). Whatever the size of the plant, its distinctive 'pea' flowers will always give it away.

The bean family's big trick is that they have domesticated nitrogen-fixing bacteria, which grow in the nodules found on the roots of many species. This makes them some of the few plants that have their own nitrogen supply. Most of the Fabaceae found in Scotland are climbers, either twining or scrambling up other plants or up supports provided by the gardener. They provide edible seeds, seed pods, leaves, flowers and in some cases roots.

Blueberry flowers, with the typical shape of the heath family

Pea flowers, showing the typical shape of the family

As my gran used to say, "The gorse is in flower: it's kissing time!" (Gorse nearly always has a few flowers, whatever the time of year.)

The mallow family

The Malvaceae or mallow family are a very important group worldwide, with members including cotton, cacao and durian. They tend to have showy flowers and you might be most familiar with them from the flower garden: hollyhocks, hibiscus, mallows and tree mallow (*Lavatera*) all belong. In the forest garden the most useful members are the mallows (*Malva*) and limes (*Tilia*). One of the most obvious characteristics of family members is a tendency to be mucilaginous or, to put it more directly, slimy. This perhaps reaches its peak in the well-known jellyish quality of okra or gumbo.

Common mallow flower

The cabbage family

The Brassicaceae or cabbage family, also known as the Cruciferae, crucifers or mustards, contains a vast number of useful plants, including all the forms of *Brassica oleracea* (cabbage, kale, Brussels sprouts, cauliflower, kohl rabi and the rest), assorted cresses and mustards, turnips, radishes, rocket, horseradish and oilseed rape. Besides their cross-shaped flowers, the greatest family resemblance is in their use of hot-tasting chemicals as a defence. The defence has backfired somewhat in the case of humans as we have developed a taste for their flavours.

Kale in full flower

Pollinators are big fans of brassica flowers

Cabbage family seedlings have distinctive double-lobed seed-leaves

Grasses have domesticated humans, making them some of the most successful plants on the planet

The grass family

The Poaceae, better known as the Gramineae or grasses, is the fifth largest family of plants. With members like rice, wheat, rye, barley, oats, sorghum, sugarcane, maize and millet they are the most important family in the world economically, but in the cool temperate home garden they sadly tend to feature more often as weeds. The exception is the subfamily of bamboos, which are simply giant grasses. Grasses owe their dominance to two clever tricks. One is a growth habit that involves pushing leaves and stems up from the base rather than growing from a growing point at the top like most plants. This makes them resistant to grazing and fire. The other is an upgrade to photosynthesis called the C4 pathway which makes them more tolerant of drought and low CO_2 conditions. These two abilities have allowed grasses to form the great prairie and steppe ecosystems of the drier parts of the world.

Borage flowers

The borage family

Many British wildflowers are members of the Boraginaceae, including comfrey, borage, green alkanet, forget-me-nots, lungworts and the oyster plant. Most use toxic chemicals called pyrrolizidine alkaloids for defence and tend to be rather hairy. This means that while many are somewhat edible it is undesirable and probably unwise to consume large quantities of any of them. The North American waterleaf is a milder-mannered exception. Most perennial members have fleshy, spreading root systems and some can become persistent weeds as a result.

One cute feature of the borage family is that many of them, like this lungwort, have flowers that change colour as they age

The knotweed family

The Polygonaceae or knotweed family includes the docks, sorrels, persicarias, rhubarb, buckwheat, bistort and the feared but very tasty Japanese knotweed. They tend to invest a lot in their roots, with spreading root systems or thick, fleshy taproots. Most use oxalic acid for defence, giving them a characteristic sour taste. Apart from buckwheat and rhubarb, few of them have become established crops, but many more are foraged around the world. They usually have small individual flowers but large, impressive flower heads, which

Persicaria flowers from the knotweed family

Allium altissimum flower

Apple mint (*Mentha suaveolens*) flower

are often edible. They may also be attractive, and many members of the family have been introduced to the flower garden.

The amaryllis family

The Amaryllidaceae or amaryllis family is packed with plants of interest to the ornamental gardener, from daffodils to agapanthus, but only one genus from one subfamily really makes it into the home garden. This is the genus *Allium*, comprising onions, garlic, leeks, chives and between 256 and 975 relatives, depending on which classification scheme you follow. They are often lumped together in gardening guides as 'the alliums', which does no justice to their diversity at all. You will find them scattered throughout the following chapters. The family's habit of early growth followed by a resting phase in a bulb serves them both in arid climates and soils where spring moisture is followed by summer drought, and on the deciduous forest floor, where spring light is followed by summer dark. This gives them a curious distribution between what otherwise seem like two very different habitats.

The alliums also share a characteristic smell, which comes from chemicals called cysteine sulfoxides. The amount of these depends on the sulphate content of the soil the plant grows in. In the rare case of sulphate-free growing conditions, all alliums completely lose their smell! Apart from this, the most obvious common features are linear (sometimes broadened) leaves – often bundled together into leek-style shoots – and flowers borne in a starburst formation on a long stem. The flowers often come wrapped in a papery bract when young, which the blooms rip open as they burst out.

The mint family

Members of the mint family or Lamiaceae typically have square stems, strong smells and small, asymmetric flowers. Some tropical members, such as teak, are trees, but cool temperate members tend to be small. They mostly feature in the forest garden as herbs (such as mint, rosemary and oregano) or weeds such as red deadnettle.

Looking down on a
young *Aralia elata*

7

Forest Garden Trees

Trees are best known in British horticulture for fruit, but the canopy layer will also yield nuts, salads, herbs, spices, seeds and shoots. The plants in this chapter range from small, shrubby trees to those that will grow into forest giants if allowed. All of them are suitable for a small home garden if pruned or trimmed, which often has the added benefit of increasing their yield. We begin with the trees of the rose family, home of most of our traditional fruits, plus a few less familiar.

Rose family

Apples
Malus domestica

ROSACEAE

Apples have been used, grown and bred by humans for a long time. There is even a theory that they were first 'domesticated' by brown bears in Central Asia. Apple seeds pass through a bear's digestive system unscathed and end up planted in a pile of rich fertiliser when the bears do in the woods what bears proverbially do in the woods. Apples that appealed to the bears' tastes were more likely to be spread and, fortunately for us, bears share our taste for large, sweet fruit.

The garden apple, *Malus domestica*, is a hybrid of several ancestral species. Apples now exist in a huge range of cultivars, each kept going as a clone by grafting onto new rootstocks to create new trees with identical characteristics. They vary in vigour

and disease resistance and in the shape, colour, sweetness, taste, texture and keeping qualities of their fruit. They are also adapted to different growing conditions and each area will have its own local cultivars.

The most important step in growing apples is choosing a cultivar. I chose badly when I first got my allotment, planting two types that do well in the northeast of Scotland but which are both early-fruiting eaters that don't keep well: 'Red Devil' and 'James Grieve'. The result was that I had a large but short-lived glut of apples around September. When choosing yours, try to go for at least one early eater but also at least one good keeper to give you apples all through the winter. Apples are often divided into 'eaters' and 'cooking apples', but this distinction is somewhat fuzzy. Apples continue to ripen in storage and a tart cooking apple can become an enjoyable eater later in the season. Equally, there is no reason not to use 'eating' apples in cooking. Avoid large national garden centres when choosing your apples as their stock can be the same across the whole country and I have seen too many cultivars offered in such centres in northeast Scotland that are not suitable for the area at all. In Scotland a large collection of local cultivars was begun by John Butterworth and is now kept and sold by Andrew Lear, trading online as Appletreeman.

As well as suitability for your conditions and fruiting qualities, you need to consider pollination. A few cultivars are self-fertile and don't need another tree to pollinate them, but most are self-sterile and need a pollination partner. In cities this is not usually a problem as there will be lots of other apples around, but in an isolated garden it can be. You then need to make sure not only that you have at least two trees but that they are good pollinators for each other. Trees can be bad pollinators because they flower at a different time to the other tree, because they are the same or a closely related cultivar (making them essentially the same tree as far as genetics are concerned), or because they are sterile triploids.

To help you match fruiting times, trees are organised into numbered or lettered 'pollination groups' that flower at the same time. The exact schema varies from place to place. Trees will pollinate if they are in the same group (flowering at the same time) or adjacent groups (overlapping). To find out if two trees are good mutual pollinators, you can find online matching tools like the one offered at Orange Pippin Trees. Trees with a long flowering time and abundant pollen, like crab apples, are good for pollinating all trees.

Finally, you also need to choose the roots of your apple. Just as there is a range of cultivars that make up the top of your tree, there is a range of rootstocks that they are grafted onto (see p.58), with varying results for the size and vigour of the tree. The following table gives the characteristics of some common rootstocks.

'Red Devil' apples

Apple blossom

Name	Size of tree
M27 (extremely dwarfing)	1.2-1.8m / 3.9-5.9ft
M9 (dwarfing)	1.8-2.5m / 5.9-8.2ft
M26 (semi-dwarfing)	2.2-3.0m / 7.2-9.8ft
MM111/M9 combination	about 3m / 9.8ft
MM106 (semi-vigorous)	2.5-4.5m / 8.2-14.7ft
MM111 (vigorous)	4.0-4.5m / 13.1-14.7ft
M25 (very vigorous)	4.5m+ / 14.7ft+

Apples like medium to light soils with plenty of organic matter and good drainage. They will grow on heavy clays or poorly-drained soils but are then more susceptible to diseases. Regular rainfall or watering is important, especially for dwarfed trees. Prolonged dry spells and sudden downpours can cause fruit to swell too fast and crack. Trees are more productive if pruned: see p.66.

Being highly-bred and usually grafted, apples suffer from a bewildering range of diseases, with names like blossom end rot, flyspeck and sooty blotch complex. Fortunately these are mostly rare or the result of poor growing conditions. The ones you might need to worry about include scab and canker.

Scab is a fungal disease which attacks the

Scab on apple leaves

leaves, fruit and twigs, causing round, brown/black spots to appear. Mild cases don't make the fruit inedible, but they do reduce its keeping time radically as they allow the fruit to dry out much faster. The fungus overwinters on fallen leaves and reinfects the tree each year, so pick off infected leaves when you see them and rake up fallen ones, putting them in a closed composter. A simpler solution is to start with cultivars that are scab-resistant (although not entirely immune), such as 'Lord Derby' or 'Red Devil'.

Canker attacks the bark, causing it first to swell, then to weep fluid and then die. If it girdles a stem then the whole twig or branch above the canker will die. It is exacerbated by heavy, wet or acid soils. The main way to deal with canker is to cut it out during pruning (p.66). Where it is not possible to cut the whole cankered shoot out, the canker can be scraped out with a knife or chisel, cutting back to clean wood.

There is much more to eating apples than raw fruit, purée and baking. They contain a lot of pectin, which helps other fruit to set in jams. They make an excellent jam along with brambles, which are available at the same time, or on their own. They are a good ingredient for fruit leather, cooked lightly before blending. I like to add a little cinnamon to mine.

They dry very easily and sweeten as they do so, and my favourite use is perhaps apple rings. To make these I take out the core using an apple corer, cut into slices 2-4mm wide and thread them onto bamboo canes. These are hung up and normally air-dry in a few days. Most varieties of apple are suitable: I actually prefer ones with a bit of bite as they lose much of their acidity as they dry.

Apples also make a good ingredient for cooked main dishes. European cuisine has got out of the habit of using fruit in 'savoury' dishes, but it is

Apple rings drying

Cairn O'Mohr cider

well worth rediscovering. My first encounter with apples used in this way was in a delicious apple curry made by Graham Bell. Ever since I have been trying apples in almost everything while they are in season. Apples have textures, rather like the 'waxy' and 'floury' kinds of potatoes; depending on the amount of starch they either keep their shape or explode into a mush while cooking. They are a staple ingredient in my winter stews and go well in soups: apple, parsnip, lentil and ginger is one of my trusty combinations.

Finally, of course, apples can be juiced, either for the plain juice or for fermenting into cider. This requires some equipment. Small electric or hand-cranked juicers take up the least space, but tend to require a fair amount of preparation and cleaning for each small batch of juice. I keep a small manual wheatgrass juicer for doing a glass of juice at a time.

Traditional apple presses take up more room and are still quite a bit of work. They are ideal for community projects which can share the press and make an enjoyable event out of the work required. At the other end of the technological spectrum, many community orchards are now investing in electric juicers which extract, pasteurise and bottle the juice. These can be available for loan or hire, either from community groups or commercial suppliers. If you live in the right area, you may also be able to trade your apples for juice or cider. In Scotland, The Cairn O'Mohr winery, run by Judith and Ron Gillies in the Carse o Gowrie, and the Clyde Cider project, run by John Hancox in Glasgow (probably not coincidentally, these are Scotland's two biggest historic apple-growing areas) operate such schemes.

Storing apples well is a challenge, but can greatly extend the apple season and the usefulness of the species. Choose cultivars with a reputation for keeping well and individual apples with unblemished skin and no bruising. Keep them so that they aren't touching each other. Some people wrap them individually in paper, but I find that the best

trade-off between simplicity, storage space and ease of inspection later on is to get the boxes and cardboard trays that supermarkets use to sell fruit in. They are normally happy to let you take them away for free as they have no further use for them. Keep them in the coolest frost-free place that you have available and you can have apples right through until April. Check up on them regularly and take out any that are starting to rot.

Pears
Pyrus species

ROSACEAE

Pears are closely related to apples, but with a flavour and texture all of their own. Three main species are grown: *Pyrus communis* is the European pear, *P. bretschneideri* is a hybrid species grown in northern China and *P. pyrifolia* or nashi pear is native to East Asia. The last species seems doomed to redundant naming: its Latin name means pear-leaved pear and 'nashi' is Japanese for pear, so its common English name means 'pear pear'. Nashi pears are juicy, sweet and fragrant – and confusingly apple-shaped.

Pears are somewhat marginal in cool temperate areas, so give them the sunniest, most sheltered niche you can find. For maritime areas, pick the

Pear blossom

earliest-ripening cultivar you can find. In continental areas the problem can be cold damaging either the cultivar or the rootstock: look out for newer, cold-hardier cultivars bred by crossing with the Asian species *P. ussuriensis*. The cultivars 'Beurré Hardy', 'Williams' and 'Jargonelle' have some resistance to frost damage to the flowers, while John Butterworth recommends the last two of these plus 'Bristol Cross' for high rainfall areas such as the west of Scotland.[1]

Fruit often doesn't ripen on the tree (and if they do, they are usually dropped and bruised), so they are best picked before the end of the season and given an extra period of ripening in a cool storage place. The window of edibility between a pear ripening and rotting is notoriously short, so keep a close eye on them. The Scottish nurseryman David Storrie writes of some pears that are 'ripe at noon, rotten by one o'clock'![2] They will, however, keep much longer in the fridge once they are ripe.

Apart from this, growing pears is similar to growing apples. They need similar soil and the same pruning regime and suffer some of the same diseases. They are grafted on to quince rootstock: 'Quince A' produces slightly larger trees of 3-4m (9.8-13ft) height and spread; 'Quince C' makes a smaller tree of 2-3m (6.5-9.8ft).

The uses of pears are similar to the uses of

Nashi pear

'Beurré Hardy' pear

Four pears: 'Williams (green)', 'D'Anjou', 'Forelle', and 'Williams (red)'

107

Getting creative with pears: pear and walnut pizza

apples too, although they don't dry well. They preserve well by bottling and are well worth experimenting with as a cooking ingredient. I'm particularly fond of them in stir-fries.

Plums
Prunus, Subgenus *Prunus*

ROSACEAE

Plums are, I think, the most luscious fruit in the cool temperate home garden. A ripe plum, bursting with juice, flavour and sweetness, is one of the great pleasures of late summer, but the right choice of cultivars will give you plums in autumn too.

There are a number of species worth growing, all of them in the *Prunus* subgenus, distinguished from cherries and bird cherries by a groove down one side of the fruit. The domesticated plum is *P. domestica*. Some subspecies of *domestica*, such as *italica* (gages) and *syriaca* (true mirabelles) are less hardy and difficult to grow in the north of Scotland. The subspecies *insititia* (damsons and bullaces) makes small, hardier trees. Their fruit tends to be acid or astringent and is mostly used for cooking or making alcohol.

'Victoria' is a favourite traditional cultivar of domestic plum, but more modern derivatives like 'Opal' and 'Excalibur' should be preferred for better disease resistance. 'Gordon Castle' is one of the few Scottish-bred cultivars still available. 'Denniston's Superb' is a hardier alternative to gages with green fruit. Of the damsons, Cox and Beaton recommend 'Farleigh Damson' as the most reliable fruiter for Scotland and 'Merryweather' for cold and exposed locations.[3]

Prunus cerasifera is thought to be the wild ancestor from which domesticated plums evolved. It is described by Andrew Lear as "Scotland's most undervalued fruit" and I completely agree. Although its native range is Western Asia and Southeast Europe it is widely naturalised in the UK and common here in Aberdeen. Its common names lead to a lot of confusion. 'Cherry plum' leads some to assume that it is a cross between cherries and plums, while 'myrobalan plum' causes misidentification with the less-hardy mirabelles (which they can also look very similar to). This confusion sometimes extends even into the nursery trade. There are a few named cultivars of cherry plum, such as 'Gypsy', 'Countess' and 'Golden Sphere'. 'Countess' is a freestone cultivar, meaning that the seed is not stuck to the flesh of the fruit but separates easily.

Feral cherry plum fruit is very variable. With some you can clearly see the reason for the common name as they are no larger than cherries; others are more plum-sized. The colour ranges from yellow to

Yellow cherry plums: the cool temperate world's apricot

Japanese plum blossom is spectacular

a speckled red through to a dark, plummy purple. Taste and texture vary too, but in general they are nice, but not strongly-flavoured for plums. They are generally poor keepers and have a habit of falling off the tree the second they are ripe (although again, this varies). This means that they are not the best eaters, although they are juicy and somewhat more-ish when munched directly off the tree.

Prunus salicina, the Japanese plum or sumomo, is a bit of an enigma. Guides generally describe it as not hardy, but a tree, which I believe to be of the cultivar 'Methley', flourishes in my garden. It is an early flowerer but seems untroubled by snow lying on the flowers. I have planted a range of other cultivars around the housing estate where I live in order to see if any will match the performance of Methley; so far none have fruited but it may be a matter of time. Most of the hard, tasteless, round plums sold in supermarkets are *salicina*, but this gives a totally unfair impression of their worth.

Commercial plums are picked unripe for transport; allowed to ripen properly on the tree they are the most richly-flavoured plum I know, with a lush and juicy texture. My tree is very heavy-yielding, producing 100kg from a medium-sized tree in one year.

All these plums are small-to-medium trees, but some others are better described as shrubs or small trees at best. The native Scottish member of the subgenus, *Prunus spinosa*, also called sloe or blackthorn, is an example. Its small, round fruit are very astringent and mostly used for making liqueurs. It suckers aggressively and if you want to make sloe gin I would suggest foraging rather than giving it a space in your forest garden.

Prunus nigra, the Canada plum or black plum, is also rather shrubby. Its fruit is somewhat sour and very juicy. Dried, it was a winter staple of First Nations in the areas where it grows naturally and it is still foraged. There are some improved cultivars,

Shotholes on Japanese plum

position. I wouldn't particularly recommend using it in forest gardens or the north of Scotland.

Plums suffer from a number of diseases, including silverleaf, plum pocket and peach leaf curl. Leaves affected by silverleaf look like they have been spray-painted with a metallic silver paint. This symptom is followed by the death of the affected branch. If you cut the branch you will see that it has a dark centre, which is the fungus invading the wood. Silverleaf can be remedied by cutting out the affected branch as soon as you see symptoms, making sure that you cut far enough down the branch that you take out all the infected wood. Much better is to avoid getting it in the first place, which is done by pruning in July, when fewer spores are present and the tree is actively growing to seal any wounds as fast as possible.

Plum pocket is caused by another fungus, *Taphrinum pruni*. *Taphrinum* is a genus of fungi that cause 'witches' brooms' in a range of trees: crazy, thick clusters of twigs caused by disrupted plant hormones. On plums it also disrupts normal fruit development, producing, as the name suggests, inedible pocket-shaped fruit with no seed. The only cure is to remove infected fruit and cut out any witches' brooms from the tree. In my experience sloes suffer most from this disease – another reason not to have them in the forest garden. Peach leaf curl is caused by a related fungus causing distorted leaves with red blisters. Infected leaves should be removed immediately and the disease can be controlled with fungicides, but for a bad case it is probably best to remove the tree and replace with a resistant cultivar.

Finally, as if these fungi weren't enough, plums can also suffer from bacterial canker, which causes 'shotholes' in the leaves (small holes which make them look like they have been attacked with a shotgun) and sunken patches in the bark that often weep resin. Good pruning practice and general tree care should avoid bacterial canker; if it develops, the infected branch should be cut out and disposed of. The resin associated with canker can

including 'Assiniboine' and 'Cheney'. It has been crossed with *P. salicina* to combine Canada plum's cold-hardiness with Japanese plum's eating qualities, giving cultivars such as 'Pembina', 'Superior' and 'Patterson Pride'.

Pollination is not usually a problem with plums. Most species will inter-pollinate and they are also planted widely as ornamentals with varieties like the dark-purple *atropurpurea*, so there are usually plenty of pollination sources around. Some cultivars are also at least partly self-fertile. The tendency of plums to mix it up means that there is lots of scope for amateur breeding of locally-adapted varieties, but only if you have lots of space! All wild species are diploid, with the normal two sets of chromosomes, while *domestica* are hexaploid, with an extra four. This means that domestic plums will not normally cross with wild ones.

Many plums produce suckers, which makes it easy to propagate own-root clones, but they can also be grafted. The two most common plum rootstocks in the UK are 'St. Julien' and 'Pixy'. St. Julien is semi-vigorous, meaning that it makes a tree smaller than one on its own roots but still of a decent size: 3-4m (9.8-13.1ft). Pixy is semi-dwarfing, producing fruit a little earlier than St. Julien but resulting in a tree that requires staking for its first few years, regular watering and a favourable

Japanese plum and shiitake mushroom pizza

Japanese plum fruit leather

Plum jam is one of the richest you can make. Cherry plum makes an especially good jam, as do the tarter varieties of domestic plum. The sweetest plums can make a rather bland jam. If you lightly sugar plums and leave them overnight you can get two products. The sugar draws juice out of the plums making a thick, sweet liquid that can be used as a cordial, and the partly-dehydrated plums can be frozen for later use. Plums also freeze well without this treatment, but they are best used from frozen as they do tend to go to mush on defrosting; for this reason it is essential to remove the stones before freezing. The most time-consuming part of any plum processing is removing the stones; it is well worth planting a freestone cultivar to save yourself this trouble. Ripe Japanese and cherry plums are soft enough that I find I can usually force the stone out of the fruit by cutting down onto its flat surface.

Plums dry well, either quartered in a dehydrator or as fruit leather. Japanese plum and ginger is my favourite fruit leather, or you can mix plum with other fruits to add sweetness to the mixture.

Cherries
Prunus, Subgenus *Cerasus*

ROSACEAE

Cherries have a lot in common with plums, including the pruning regime and many of their diseases. They are less versatile in the kitchen, but nothing beats eating them fresh and spilling out the stones on a summer's day. Cherry varieties are derived from either the native Scottish species (*P. avium*) for sweet cherries like 'Stella' or the non-native European sour cherry (*P. cerasus*) for cooking cherries like 'Morello'.

Cherries like a rich fertile soil and a sunny spot. Beyond this there are two main challenges to growing them. The first is that the birds like them even more than we do, and trees can be stripped even before they are ripe. One solution here is to

also be produced by plums for a range of causes. It's an ill wind: one friend tells me that he successfully made home-grown gummi bears (plummi bears?) by combining the fruit and the resin!

My favourite use of plums is simply to eat the fresh fruit raw. Birds and wasps also appreciate their sweetness, so it can be advisable to pick fruit a few days before full ripeness and allow it to finish ripening on a sunny windowsill. Japanese plum is so sweet and melting that I use it wherever I would use tomatoes, including for salads, sauces and pizzas. All plums go well in a stir-fry, or simmer with onion, garlic, ginger, chilli, and soy sauce for a Chinese plum sauce.

Cherry's ice-white blossom normally appears in early May

grow on a dwarfing rootstock such as 'Gisela 5' or train against a wall and net the trees.

The other challenge is that most cultivars are not really bred for Scotland (although Willie Duncan recommends 'Lapins' for the east of Scotland and Cox and Beaton[4] mention 'Regina',

Cherry fruit

'Summer Sun' and the Canadian cultivar 'Sunburst' as especially hardy) and most commercially-grown cherries in Scotland are raised under cover. I am certain that breeding efforts could produce better cherries for our climate. In the Speyside town of Aviemore, an accidental breeding experiment seems to have taken place, with a large housing development having been planted up with what are evidently cherry trees grown from the open-pollinated seeds of some sweet variety, perhaps from the seeds left over from someone's bowl of cherries. The results vary considerably in taste, size and colour, but the best ones reliably produce large, rich fruit in abundance every year. I have noticed that such well-adapted wildling trees are usually less bothered by birds than commercial ones, so this could be a way of killing (or preferably deterring) two birds with one (cherry) stone.

Bird cherries
Prunus, Subgenus *Padus*

ROSACEAE

The third subgenus in *Prunus* includes the native Scottish bird cherry (*P. padus*), a small tree of wet ground found across Eurasia. In East Asia there is *P. mackii* in Amur and Manchuria and *P. grayana* in Japan and China, and in North America *P. virginiana* is known as chokecherry. The common names do not inspire confidence and the unprocessed fruit of most trees in this subgenus are indeed astringent and poisonous, with high levels of tannins and a glycoside called amygdalin which releases hydrogen cyanide when the flesh is crushed.

Fortunately, the levels of these substances vary between individual trees and they can be reduced or eliminated by processing. I make a habit of trying any ripe bird cherries that I can find. Some are better than others, but I have never found any that I would really call edible. I also can't find any reliable records of historical use of unprocessed bird cherries, although Gordon Hillman writes that some people find the taste pleasant.[5] In North America, however, the picture is very different. Chokecherry was widely used in pre-Columbian times and was the most important fruit in the diet of some nations. Sam Thayer describes eating fruit directly from the tree.[6] He says that some trees are noticeably less astringent than others; my plea to foragers is to propagate and share these trees. A superior bird cherry would be easy to propagate as it suckers very freely.

The fruit can also be made edible by processing. People from the Amur Valley in the Russian Far East (with bird cherries)[7] to British Columbia in western Canada (with chokecherries)[8] have followed the same procedure. The whole fruit, stones and all, are pounded until the stones are thoroughly broken up, then the mixture is spread thinly and dried until the almond aroma of the amygdalin disappears. This works because the shells of the seeds in these species are relatively thin. It has

the virtue of releasing the nutritional value of the seeds, which are too small to extract otherwise, and has been a staple food for a number of groups, but most people would probably not enjoy the grittiness of the result. *None of these species should ever be eaten fresh with the stone inside.*

Chokecherry juice can be extracted by boiling the fruit and straining, and then used to make syrup or jelly. Thayer recommends boiling the fruit with water that has had apples boiled in it, to both sweeten and thicken the jelly. The flesh can also be removed by putting the fruit through a mill or a colander, to make jam or pies. Thayer reports

Bird cherry carries its blossom on distinctive flower spikes

Chokecherry in Parker River National Wildlife Refuge, Massachusetts

that the best use of all is fruit leather. The astringency can be reduced by picking very ripe fruit and leaving it in the fridge for a couple of days, or even freezing and thawing, before removing the flesh.

Juneberries
Amelanchier species

ROSACEAE

Imagine that you could grow blueberries on a tree! With juneberries, you more or less can. *Amelanchier* species are widespread across the northern hemisphere. North America has the most, including *A. alnifolia*, *canadensis*, *laevis*, *lamarkii* and *stolonifera*. Asia has *A. asiatica* and *sinica*. Europe draws the short straw with only *A. ovalis*, the snowy mespilus, generally considered to be a poor eater. Scotland has no native species.

The common name juneberry comes from their fruiting month. They might be better called julyberries in Aberdeen, but they are still one of the earliest fruit around. Other common names include serviceberry and shadbush, the latter from the fish that sometimes runs the rivers at the same time as the trees are in blossom.

Amelanchier alnifolia in Cruickshank Botanic Garden, Aberdeen

Berries on the same tree

A. lamarkii autumn colours

Amelanchier lamarkii

Juneberries are small, sweet and delicious, like blueberry but with a nutty taste from the seeds. I like them best simply eaten off the tree, but they also dry well to make june raisins. They are borne early enough that they will air-dry on the tree even in Scotland if the birds don't get them. This, however, is usually the problem. The early, sweet berries are irresistible to birds and if you don't like sharing you will have to net your trees. Fortunately, most species are small trees, and two of the best, saskatoon (*alnifolia*) and Quebec berry (*stolonifera*), are large and small shrubs respectively. Both of these sucker from runners, making them easy to propagate.

Juneberries are attractive enough that many have found their way into the garden trade as ornamentals, making them quite easy to get hold of. There are even some garden hybrids such as *Amelanchier* x *grandiflora*, the fruit of which is reported to vary from delicious to dull. All the species mentioned above are very hardy. They are not especially fussy about soil conditions and are not much troubled by diseases. Saskatoon will take a bit of shade but in general a sunny spot is best.

Hawthorns
Crataegus species

ROSACEAE

The list of edible hawthorns is long indeed, but it only very marginally includes the native Scottish hawthorn, *C. monogyna*. Most are small, thorny trees, although *C. tanacetifolia*, the tansy-leaved thorn, can grow to 10m (32.8ft). The fruits consist of a large central seed (which is often in fact several seeds stuck together) with a fleshy covering. The quality of the flesh varies from thin, dry and astringent, to juicy and delicious. Nancy Turner reports that red and black haws (*C. columbiana* and *douglasii*) were used by various First Nations in British Columbia, although they don't seem to have rated them very highly.[9] *C. pinnatifida* (Chinese hawthorn) is used in China and Korea to make jams, jellies, juices, alcoholic beverages, and other drinks.[10]

My problem with the genus as a whole is that even the best I have tasted are essentially small, very seedy apples. It is hard to know why they should be given a place that could instead be used for an apple tree. On the other hand, Ken Fern says of *C. schraderana*, which I haven't tried, that it has one of the most delicious fruits he has ever eaten from a plant of the temperate zone, and that he

Preserved Chinese hawthorns

115

would far rather eat this fruit than a strawberry.[11] Given the number of fruits that Ken Fern has tried, this is high praise indeed.

Hawthorns are quite susceptible to fireblight and can infect other rose-family trees if they are not removed quickly. Their blossom (or 'may') is a spectacular sight, but don't be tempted to take too deep a sniff. They attract flies for pollination by emitting a smell of rotting flesh! Exotic hawthorns are usually grafted onto native hawthorn rootstock in the UK. They are mostly not easy to come by; Martin Crawford at the Agroforestry Research Trust produces a small number every year.

Quince
Cydonia oblonga

ROSACEAE

Humans have been enjoying quince for a very long time: it was grown in Mesopotamia over 4,000 years ago and was known to the Akkadians. It's a rather reluctant immigrant to Scotland. It is hardy, and grows into a tree 3-4m (9.8-13.1ft) tall, but only fruits well if given a sheltered place on a south-facing wall. Even then, fruits do not ripen fully and are only used in cooking. The main reason to grow such an unpromising plant is that the fruit, even unripe, is very aromatic and can be used to flavour apple sauce, jams and pies. Be careful not to store them with apples or pears as they can also flavour the raw fruit in this way. In Spain a quince paste called membrillo is very popular. Thankfully, quinces are self-fertile, so you only need one. Nick Dunne recommends the cultivars 'Meech's Prolific' and 'Serbian Gold' for Scotland.[12]

Hawthorn flowers: look but don't sniff!

Assorted quince varieties from the germplasm collection at the USDA-ARS National Clonal Germplasm Repository in Corvallis, Oregon

Aralia family

The trees in the Araliaceae are much less familiar in the West than the Rosaceae. I first learned about many of them in a delightful article in *Economic Botany* by David Brussell,[13] who taught ethnobotany and ecology in Japan for seven years and combined an academic interest with an obvious personal passion for wild foods. In all cases it is the young shoots, recently emerged from the buds, that are eaten. The genera have been extensively rearranged, so check for synonyms if you are trying to get hold of them. The only source of many of them in the UK is Crûg Farm Plants in Wales.

Harigiri leaves

Harigiri
Kalopanax septemlobus

ARALIACEAE

Harigiri, a handsome tree with maple-like leaves, grows to 30m (90ft), so it may require some pruning to fit in your garden! It is very hardy and overwinters easily in my garden. The leaf bundles are a little bitter, and are cooked either as tempura or by blanching (immersing in boiling water a few times until the water doesn't go green and the bitterness is removed). Once blanched they can be dipped in sauce and eaten.

Blanched harigiri shoots in a dressing of soy sauce, sesame oil and lemon juice

Koshiabura
Chengiopanax sciadophylloides

ARALIACEAE

Koshiabura, also known as gonzetsu, is regarded as the 'queen of the sansai' and was one of David Brussell's favourites. The shoots can be cooked as tempura, sautéed or chopped into soups. Unfortunately my own tree died and I only got to try the shoots once, but I can confirm that they are tasty. It is hardy down to -10 to -15°C (14 to 5°F), so it should work in many cool temperate areas. David Brussell describes it as a sizeable forest tree. Unlike most of its relatives it is not spiny.

Koshiabura

Aralia elata: any devil would be proud to have a walking stick like this (see also p.102)

Devil's walking sticks
Aralia elata and *spinosa*

ARALIACEAE

Aralia elata is a small tree native to China, Korea and Japan. It is very striking, with large, doubly-pinnate leaves, brick-red fall colour, sprays of white flowers and gnarled, knobbled, spiny branches. The spring shoots (known as *taranome* in Japan and *dareup* in Korea), picked before the prickles harden, are a popular sansai, often cooked as tempura. There is also a spineless form which must look less impressive in the garden but can be eaten raw.

Aralia spinosa is a closely-related tree from North America. Its fearsome-looking stems give it one of its English common names: devil's walking

stick. I prefer this name to the alternatives, angelica tree, prickly ash, prickly elder and Hercules' club, all of which invite confusion with unrelated plants. Since *A. elata* suffers a similar confusion of English names, Stephen Barstow suggests 'Oni's walking stick', after the appropriately gnarled and spiny troll of Japanese folklore.

A. spinosa has some use as a pot herb in North American foraging. David Brussell, returning from Japan and missing taranome tempura, tried some and found that it tasted very similar. He also mentions a Native American ancestor who used to make a skin salve with the leaves, stem and bark.

Both trees are woodland edge species, able to take some shade and sometimes forming thickets by throwing up suckers. Oni's walking stick is somewhat hardier and better for more northern gardens (it does well in mine), but it is considered invasive in some US states and the native tree should be preferred there.

Hawk's claw
Gamblea innovans

ARALIACEAE

David Brussell also mentions hawk's claw (*takanotsume*), also known as *imonoki*, and gives brief recipes. "The young shoots are set in a pan of boiling water for 2-3 min., then removed, and mixed with sweet miso and sesame seeds to make *aemono* (a popular dish consisting of briefly boiled greens of various sorts stirred together with miso and sesame seeds). The young shoots are put in boiling water for 2-3 min., then taken out and sprinkled with soy sauce to make a dish called *ohitashi*."

I haven't been able to get hold of this species or find out very much about it in the English literature. Plants For A Future dismiss it as famine food, but Japanese websites tell a different story, describing it as, for instance, "delicious when made into tempura".[14] It is a mountain tree growing on the northern island of Hokkaido (amongst others), so likely to be hardy. It grows to around 6m (19.6ft).

Other families

Elderberries
Sambucus species

ADOXACEAE

Elderflowers have the scent of summer: a 'muscatel' fragrance that can be used to flavour jams, jellies, puddings, cordial, wine, tea, gin and more. Flowers need to be picked on a warm, dry day when they are releasing their scent (just sniff to test). One of the simplest ways to use them is to steep a few flower heads in water overnight with a little sugar and lemon juice to taste. The resulting infusion is delicious, and if left for a while it will ferment with the elderflower's natural yeasts (or add a little baker's yeast if you're impatient) into elderflower champagne. If you can resist drinking all of this, allowing it to oxidise produces an exquisite elderflower vinegar.

For a longer-lasting supply of elder flavour, make a cordial by pouring a litre (34 fl.oz) of boiling water and 100ml (3.4 fl.oz) of lemon juice (about two lemons if using fresh lemons – add the zest too if you like it) over about 20 flower heads and leave to soak overnight. Then strain out the solids, add about 700g (24oz) of sugar, heat to dissolve, simmer for a couple of minutes and bottle. The cordial can be diluted for drinking or used to add elderflower flavour to drinks and puddings.

One of my favourite uses is for rhubarb and elderflower jam. No liquid is required. Chunks of rhubarb are layered with sugar in a tub. Halfway up, I put the flower heads, face down, and then continue layering. Overnight, the sugar draws the juice out of the rhubarb and the flowers steep in it. The next day I cook the rhubarb as jam; the flowers are taken out and steeped in water, producing an elderflower cordial or champagne with the added sourness of the rhubarb. Two for the price of one! Three kilos (6.6lb) of rhubarb, 1.5kg (3.3lb) of sugar, 10 flowers and the juice of two lemons makes about 10 jars of jam.

If none of that appeals to you, you can fritter the flowers instead, by dipping them in batter and then frying them (do this in shallow oil in a frying pan, not deep fat, in order to retain the flat shape of the flowers). The flowers taste delicious but the stems taste horrible, so you need to use the stem as a handle while you nibble off the battered flowers.

If you remember to leave some flowers, the next crop is the purple-black berries. Many people find them laxative or sickening raw (with the seeds as the likely culprit), so they are either cooked or juiced. One challenge is removing the berries from the unpleasant-tasting stems. How thorough you need to be depends on what use you have planned

Elderflowers

Elderflowers steeping in rhubarb and sugar

Elders are usually small trees, but this one in Cromarty has reached over eight metres

Elderberries

for them. One popular way of removing berries from stems is stripping them off with the tines of a fork. Sam Thayer suggests putting them in the freezer (loosely stacked). When they are taken out again the frozen berries can be separated quickly and easily by rubbing them between your hands.

The berries can then be made into a jam or juiced by mashing, boiling and straining through a jelly bag. Juice makes the best wine there is (in my opinion), the best fruit vinegar there is (in Miles Irving's opinion),[15] jelly, syrup or plain juice, useful for mixers or as a tonic. Elderberry juice has been shown to have antiviral properties, a good excuse for a glass of elderberry wine when you're feeling down.

Forager-herbalist Mandy Oliver takes a different route to elderberry vinegar. She steeps 350g (12oz) of fruit in 500ml (17 fl.oz) of white wine vinegar

Cairn O'Mohr elderberry wine

for 3-5 days, stirring occasionally, and then strains. She adds 350g (12oz) of sugar per 260ml (9 fl.oz) liquid, boils for 10 minutes and bottles. She regards it as a better alternative to balsamic vinegar.[16]

There is one more, lesser-known, harvest from elder trees. A number of eighteenth and nineteenth century recipe books[17] give recipes for pickling peeled elder shoots, often pitching them as a home-grown substitute for exotic bamboo shoots. I have tried this and find them very enjoyable. You need the long, straight shoots that elder is in the habit of throwing up (especially where they have been cut or coppiced), picked while they are still soft and bendable. Only pick nice fat ones as smaller ones are not worth the bother. The shoots need to be stripped of the bitter, toxic skin, soaked for a day in strong brine and then pickled in brine or vinegar. All parts of the elder apart from the flowers, fruit and the pith of these young shoots are poisonous.

The few cultivars of elder have mostly been selected for ornamental rather than edible use, such as 'Black Beauty' and 'Black Lace'. They can be used just like wild elders and the 'black' varieties make a nice pink elderflower champagne. Cairn O'Mohr winery keeps an orchard of selected cultivars found in the wild, with names like 'Car Door' (so heavy-fruiting that the sound of the berries hitting

the bottom of a collecting bucket was mistaken for a car door slamming). All foragers agree that some trees have nicer flowers and berries than others, so there is scope for this process to continue.

North America has many elders. *S. canadensis*, the American elder, is very similar to *S. nigra* and used in the same ways. *S. caerulea*, the blue elder, is widely considered the best of the bunch. They grow larger than American elder – up to 12m (39.3ft). The berries are also larger than those of the black elders. Sam Thayer says they have "a pleasant, fruity, grape-like sourness" and greatly enjoys eating them raw.[18] They make the best jelly due to their acidity and are also used in pies. They are hardy, growing on the west coast up to southern British Columbia.

There are also several red-berried elder species, one of which, *S. racemosa*, has become somewhat invasive in Scotland. They are much inferior to the blue and black elders and, while they are sometimes foraged, they are not worth planting in a home garden.

All elders will take some shade, but fruiting is then reduced.

Hazels
Corylus species

BETULACEAE

The hazel is the only nut tree that is truly suited to cool temperate regions. As such, people have been making use of its oily, protein-rich gifts for a very long time. We have evidence of hazelnut cultivation in China dating back over 5,000 years,[19] while in Scotland a sizeable nut processing operation has been uncovered on the island of Colonsay, dating back to almost 8,000 years ago.[20] The Colonsay hunter-gatherers appear to have been roasting the nuts, which makes them tastier, more digestible and longer-keeping. I do mine on top of my stove.

The northern European hazel is *Corylus avellana*. In Asia there are *C. sieboldiana* (Asian beaked hazel) and *C. heterophylla* (Asian hazel), and in

Hazel nuts in Cromarty Courthouse orchard

Hazel harvest

North America *C. americana* (American hazel) and *C. cornuta* (American beaked hazel). There is also *C. maxima*, the Turkish hazel or filbert, which has larger nuts but is less hardy. Hazel cultivars can be selections of *avellana* or crosses with filberts. Nick Dunn's filbert recommendation for Scotland is 'Cosford'[21] and Andrew Lear sells the cobnut 'Webb's Prize'. I have found many larger-seeded trees while out foraging; fortunately hazels are easy to propagate by layering or digging up suckers so if you find a good, locally adapted tree you can take a copy of it for your garden. This might well be better than buying a cultivar which has been bred for conditions further south. Hazels are wind pollinated and not self-fertile, so you will get more nuts with more than one cultivar, planted close together. Because hazels are grown on their own roots there is no need to worry about rootstocks.

As well as the nuts, hazels were traditionally grown for coppice products like poles. If you manage for nuts, you can harvest the occasional pole as trees have a habit of sending up long, straight stems even when well established, but if you coppice you won't get nuts. Hazels will grow in a fair bit of shade and hazel coppice was traditionally grown under oak 'standards', but for nut production it is best to plant in full sun.

Hazels tolerate many soils, but their favourite is a light, sandy, well-drained soil close to neutral pH. They like a good mulch, but over-feeding them leads to lots of woody growth and fewer nuts. Nut production can be increased by pruning – see p.66.

Pea trees
Caragana species

FABACEAE

The Siberian pea tree (*Caragana arborescens*) growing in my garden looks very healthy and bears pretty yellow flowers, but unfortunately is yet to set seed. It may need a pollination partner or it may just prefer the hotter summers of a more continental climate. *Caragana* is the only genus of trees in the bean family with edible seeds that grows well in a cold climate, so it has the potential to fill an important niche in the forest garden. Like most of the family, it fixes its own nitrogen and can grow on poor soils.

Ken Fern reports that the raw seeds taste pea-like, although it's not really certain that it's wise to eat them raw in any quantity. Plants For A Future quotes a number of sources saying that the green pods are edible, but these all seem to be old books quoting each other from an unknown original source, so I wouldn't rely on it.[22]

The cooked seeds are a better bet. Aaron Parker describes the taste as "like a cross between a lentil and a walnut".[23] Melissa Hoffman makes miso and tempeh from them and describes the miso (made after boiling in three changes of water) as "ambrosial". The tricky part is not so much the cooking as the harvesting, as the pods are small and crack open once the seeds are ripe. The preferred method seems to be to spread a sheet under the trees and give them a shake when they are ready to shed.[24] Individual trees have been reported with non-shattering pods and larger seeds and enthusiasts are working on propagating these, so these problems may be overcome in the future.

There are several other cold-hardy *Caragana*, including *C. boisii* and *C. fruticosa*.

Pea tree in flower

Cooked *Caragana arborescens* seeds

Walnuts and hickories
Juglans and *Carya* species

JUGLANDACEAE

The well-known walnut (*Juglans regia*) is not a tree of the cool temperate zone. I have seen one fruiting as far north as Perth, but at this latitude the best you will get is green walnuts for pickling. There are, however, two hardier relatives that would be worth experimenting with. The Japanese walnut (*J. ailanthifolia*) grows as far north even as Sakhalin, the Russian island at the northern end of the Japanese archipelago. The heartnut, a variety of Japanese walnut called *cordiformis*, has sweeter, heart-shaped nuts than the species. Their big drawback is they are notoriously difficult to shell.

Then there is *Juglans cineria*, the butternut. It is easier to crack than the heartnut and has a good flavour, but it is becoming rare in its native North America due to a fungal disease that heartnuts are immune to. With this spread of desirable characteristics across two species, it is not surprising that people have tried crossing them, and the result is called the buartnut. Cultivars like 'Mitchell', developed in Ontario, are now available.

I have planted a buartnut 'Mitchell's seedling' (i.e. an open pollinated offspring of 'Mitchell') and a butternut 'Bear Creek seedling' in the

Buartnut 'Mitchell's seedling' leaves

neighbourhood where I live. Both are growing well, but only time will reveal the full potential of them and other selections for the north.

The hickories are a large group of nut trees, closely related to the walnuts. For now, they don't do well in the north of Britain, but they may shift their range northwards with climate change. *Carya illionoiensis*, *laciniosa* and *ovata* are the hardiest species.

Blue sausage fruit
Decaisnea insignis

LARDIZABALACEAE

To be honest, this plant is more of a novelty than a serious food crop, but what a novelty! It is a small tree that bears bean-like pods (also giving it the name 'blue bean') with a squidgy texture and striking metallic blue colour. The pods are filled with shiny black seeds which are covered in a thin pulp that reminds me of lychees or chocolate seeds.

Blue sausage fruit

Eating it involves spitting out a lot of seeds, so it is best done outside. Not a lot of calories, but fun for children of all ages.

Charma
Hippophae salicifolia

ELEAGANACEAE

Willow-leaved sea buckthorn is a tree-sized version of the shrub sea buckthorn. It has similar fruit: small and orange, with a sour kick and lots of vitamin C. It makes an attractive tree, with narrow, silvery leaves. It can grow up to 15m (49.2ft) but tends to be smaller in cooler climates. It is dioecious, supposedly needing male and female trees for pollination, but a solitary female tree I grew in a park in Aberdeen bore fruit, so either it has a degree of self-fertility or it was being pollinated by nearby sea buckthorn. The normal English name is quite a mouthful, so I prefer the charming name of charma, a name for the berry in its native Himalayas.

Bay tree
Laurus nobilis

LAURACEAE

The bay tree, the source of bay leaves, is hardier than its Mediterranean origins and appearance suggest. The leaves gain in intensity with a few days' drying, after which they should be stored in an airtight container. It is one of those plants that produces more than an individual gardener will ever need in their kitchen, but you can cut it like a hedge to reduce its space demands or supply your whole community.

Succulent lime leaves

Lindens
Tilia species

MALVACEAE

Tilia species are known as limes, lindens or bass-woods, but in my household they are most commonly called salad trees. Two species are native to Scotland: small-leaved lime (*T. cordata*) and large-leaved lime (*T. platyphyllos*). Of the two I prefer the small-leaved, because the leaves of *platyphyllos*, although larger, tend to be rather hairy and have an unpleasant mouth-feel. The two also hybridise, giving *Tilia* x *europaea*. Hybrid limes can be any mix of parental characteristics, from the best of both worlds to the worst, so it is worth keeping a look out for superior trees. They are widely planted as street and park trees, so that shouldn't be difficult. A good tree could be propagated by grafting, from suckers or by rooting cuttings taken in summer with rooting hormone.

Many trees have edible leaves at some point, but limes are the only ones I know that will give a continual supply of salad-quality leaves for most of the growing season. The trick to getting this is to pick whole shoot tips instead of individual leaves.

Tree bumblebee (*Bombus hypnorum*) on a linden flower

125

This forces the tree to grow out new buds, with new tender growth, so the more you pick the more you get. You'll need to cut your lime like a hedge to do this so that you are picking most of the shoots.

Like most of the mallow family, lime leaves have a slight mucilaginous quality. I wouldn't make a salad of pure lime leaves, but they provide a mild-tasting base to mix with other, stronger leaves.

The only drawback to picking so many leaves is that you will severely reduce the amount of the linden's other sought-after part: the flowers. These are also edible when young and can be made into a sweet, aromatic tea.

North American *Tilia* species are known as basswoods: *T. americana*, the American basswood, is the cool temperate species. The leaves are just as nice as the European species'. The soft, sweet cambium, the layer directly under the bark, was used in the past as fibre for ropes and cloth ('bast', hence the name). It can also be eaten while the sap is running, although if you harvest too much you won't have a tree any more.

Chinese mahogany
Toona sinensis

MELIACEAE

Chinese mahogany is the only 'true mahogany'* able to grow in northern parts. It does indeed produce beautiful timber if you have 80 years to wait. In the meantime, content yourself with the young shoots and leaves, which have an intriguing taste, putting me in mind of onions and spices. They can be boiled, stir-fried or ground into a distinctively flavoured paste used in Chinese cooking. In my garden, the first shoots sometimes get frosted, but it has always come back quite successfully from buds further down the stem and the dormant

---

* *Swietenia macrophylla* is known as 'genuine mahogany'. Other trees in the Meliaceae, including Toona, are 'true mahoganies'.

Fresh and fully expanded leaves of *Toona sinensis*

tree is hardy to about -25°C (-13°F). Unfortunately I might get into trouble with the City Council if I let my tree grow to its full 20m (65ft) height, so I manage it as a hedge which makes picking easier. Chinese mahogany is also known as red toon or Chinese cedar (botanically inexplicable, but probably based on its use, like cedar, as an incense).

Pines
Pinus species

PINACEAE

The Scots pine (*Pinus sylvestris*) is well known as the mainstay of the Caledonian forests of the Scottish Highlands and as a timber tree, but less known as a source of two edible products. The first are the male shoots, bearing the immature pollen cones: a cluster of pale yellow globes that almost entirely obscure the young shoot. At this stage I eat the entire shoot, which has an amazing resiny, nutty flavour. They hold together well in cooking and look as well as taste good. My favourite uses for them are stir-fries and pasta sauces. The short harvest period is offset by the fact that they freeze very well and keep their shape and flavour on thawing. I have tried shoots from other species of pine, but none seem to compare with the Scots. If you miss the harvest period for the immature cones, don't worry, you can harvest the pollen itself, which makes a flour with a similar resiny/nutty flavour.

Despite the British name, 'Scots' pine is found across temperate Eurasia, with one of the widest distributions of any tree. It is a big tree, too big for a small garden but an attractive addition to a larger one. They grow best on well-drained, not-too-rich, acid soils.

The other reason to plant a pine in a forest garden is for the large, oily seeds of some species: the pine kernels or nuts. Most pine nuts sold in shops nowadays are from the Korean pine, *Pinus koraiensis*, which is a hardy tree worth trying in cool areas if you have the space. Other Eurasian nut pines include the closely related stone pine (*P. cembra*)

Scots pine shoots

Collecting pine pollen

Making pesto with pine pollen cones

127

Xanthoxylum bungeanum leaves and fruit

and Siberian pine (*P. sibirica*). Stone pine grows in the Alps and Carpathians, Siberian pine in Siberia and Mongolia. I have seen a stone pine growing in Aberdeen, but not producing seeds. This may be because all the pines need a pollination partner, so you'll have to plant at least two of any kind you try.

In North America, Ponderosa pine (*P. ponderosa*) and whitebark pine (*P. albicaulis*) have been used for seeds. Nancy Turner quotes a Secwepemc woman on Ponderosa pine: "Once you start eating them you can't quit."[25] Ponderosa pine comes in a range of subspecies, of which subspecies *ponderosa* is the most northerly. Ponderosa pine, whitebark pine and lodgepole pine (*P. contorta*) all have edible cambium, like limes.[26]

Sichuan pepper and prickly ashes
Zanthoxylum species

RUTACEAE

Sichuan pepper, although sharing a name with black pepper and chilli pepper, has a taste and effect all of its own, described by Harold McGee in *On Food and Cooking* as producing "a strange, tingling, buzzing, numbing sensation that is something like the effect of carbonated drinks or of a mild electric current".[27] It is the fruit or seed pod of *Zanthoxylum bungeanum*, a small tree in the same family as the citrus fruits.

The Sichuan pepper tree is a promising candidate for the forest garden. A small grove flourishes in the Cruickshank Botanic Garden in Aberdeen and Ken Fern reports that a plant in the Cambridge Botanical Gardens grows well and fruits in deep shade.[28] Plants are dioecious, with separate male and female plants, so you need at least two plants

The spices of the mountain pepper tree are well guarded

to be sure of producing viable seed, but one sup-plier[29] states that most plants will fruit even if not fertilised, and it is the pod rather than the seed that is used. At worst, the bark and leaves share the same pungent aroma as the fruit.

The naming of this plant can be confusing. Most suppliers distinguish between *Z. simulans* (Sichuan pepper) and *Z. piperitum* (Japanese pepper), but botanists have now decided that these are the same species as the northern Sichuan pepper, or *Z. bungeanum*. North America has its own mem-bers of the genus, the most northerly of which are *Z. americanum* and *Z. armatum*, the northern and winged prickly ashes.

Sichuan pepper fruits and husks

Flowering quince in a
walled garden in Aberdeen

8

Shrubs, Canes,
Large Herbaceous Perennials
and Climbers

Below the trees, we have a rather mixed bag of species that are too small to be called trees, but too large to belong in the ground layer. Some are woody shrubs, some die down and leap back to considerable height every year and others climb up trees or supports.

Rose family

The rose family again provides us with a selection of useful species, most of them in the genus *Rubus* or, to their friends, the berries. Some berries have an unusual genetic trick: they are apomictic, meaning that they can produce seeds which will grow into exact clones of themselves. Apomictic groups are hard to separate sensibly into species; their ability to dispense with sex for generations muddies the idea of a gene pool that our usual species concept relies on.

Add to that the ease with which different species cross, and the berries are a taxonomist's nightmare but a plant-breeder's dream. Popular cultivars can contain genetic contributions from many different species. This means that in practice you can ignore much of the complexity. You don't have to worry about acquiring the balloon berry, *R. illecebrosus*, with its giant but insipid fruit. The beneficial characteristics have already been bred

into other raspberry cultivars, with the boring taste left behind. Interesting crosses can also take place in your own garden, so you can create and easily maintain unique strains adapted to your garden.

The berries mostly evolved as woodland-edge plants so they are very much at home in a forest garden. They will bear a little shade but don't fruit well in heavy shade. Most berries have a two-year cycle for their stems. In the first year a stem (called a primocane) will grow up from the roots. In the second year it is called a floricane; it will become woody and branch, produce fruit and then die, while new one-year stems grow up around it. Dead floricanes are usually removed to prevent disease and reduce crowding. Because berry canes come up so thickly there isn't much point growing anything you want access to underneath them, but they make a good place to lay woody mulch from other plants.

Brambles
Rubus, Subgenus *Rubus*

ROSACEAE

The brambles or blackberries (*R. fruticosus* agg. and their relatives) pose a challenge in the home garden. They have the ability to root at the end of their long, thorny stems, a trick they use in the wild

Bramble flowers

Wild brambles

Making connections through foraging. I met this lady picking brambles in the local park that I used to manage, where we actively managed the wild brambles to encourage fruit. We got talking and I discovered that she was Italian, having come to Scotland as a war bride many years ago.

to invade open space at dizzying rates. Sped-up footage of growing brambles looks eerily animal as the briars probe, grab on to a suitable spot of earth and lunge forward again. They will happily do the same in your garden, creating a tangle of barbed-wire trip wires.

There are several ways to tame the brambles. The first is to choose a thornless cultivar, such as 'Loch Ness' or 'Oregon Thornless'. Cultivars also have larger fruits and heavier crops than the wild species, but I often find the taste disappointing. The second is to tie them to supports, such as a fence, wires on a wall or a purpose-built frame. This lifts the briars off the ground, making picking

easier and preventing the tips from touching the ground. The third is to exploit their growth habit; they arch up from the roots, then over and back down to the ground. Cutting all first-year stems at the top of their arc in August produces a shrubby clump. If old canes are left then this clump will increasingly self-support over time.

The reward for all this is a crop of blackberries, with a deliciously complex flavour as dark as the fruit itself. They mix well with apples or pears: in compote, in crumbles and pies, and in jam. Blackberries freeze well for later use.

Raspberries
Rubus, Subgenus *Idaeobatus*

ROSACEAE

Species in the subgenus *Idaeobatus* tend to grow with more upright, better-behaved canes than the brambles; broadly we can call them the raspberries. These include the ubiquitous red raspberry (*R. idaeus*), the eastern and western black raspberries (*R. occidentalis* and *leucodermis*) of North America, the Korean raspberry (*R. crataegifolius*) and the massive-fruited balloon berry that I mentioned earlier.

Scotland is ideal territory for raspberries and a major centre for their breeding. This is reflected in the names of cultivars (which increasingly contain contributions from many different raspberry species), such as 'Glen Moy' and 'Glen Ample'. They are unusual in that breeders have managed to achieve a considerable increase in the size of fruit with little loss of flavour. Raspberries in the wild also vary enormously and it can be worth taking a good wild strain into your garden.

With a range of different cultivars in your garden you can enjoy raspberries from June to December. In my garden the wild types are usually the first, followed by 'summer fruiting' cultivars. Last up are 'autumn fruiting' plants like 'Autumn Bliss' and 'Joan J'. These vary from the *Rubus* norm in fruiting on year-one canes – see Chapter 4 for the pruning regimes for different types of raspberry.

You can also choose between a range of colours. Beyond aesthetic appeal, yellow-fruited raspberries, like the summer-fruiting 'Amber' or the autumn-fruiting 'Allgold', are less attractive to birds. I have a wild strain that I call 'Sunset' that turns from green to yellow to salmon orange as it ripens, making it simple to pick just the ripest, sweetest berries. Purple and black cultivars are either selections of the black raspberry species, like 'Bristol', 'Jewel' and 'Black Hawk', or crosses

Raspberry flowers

Picking summer fruit

Colour change during ripening in 'Sunset' raspberries

between red and black, like 'Glen Coe' or 'Malling Passion'. The growth habit of black raspberries and their hybrids tends towards that of the brambles, with more trailing stems.

Aside from eating them fresh off the plant, raspberries are wonderful in all sorts of summer puddings, in jam and for wine. They freeze extremely well, and I enjoy them straight out of the freezer topped with cream (which promptly freezes on to the berries). A lesser-known product is the peeled shoots, especially of the thicker stemmed varieties of black raspberry. For this they need to be picked while the year-one stems are newly growing.

There are a couple of outliers in the subgenus: Japanese wineberry (*R. phoenicolasius*) and the Canadian raspberry or salmonberry (*R. spectabilis*). Wineberry is an extraordinarily sticky plant, due to red, glandular hairs that cover most parts of it – but the fruit is excellent. It grows more like a bramble than a raspberry, although not so vigorously. Try to get seed or a plant from a local supplier as they are less hardy than other raspberries and a local strain is likely to be better.

Salmonberries, on the other hand, should be avoided in my opinion, despite often being offered by suppliers of unusual fruit. The fruit is insipid and yields are poor, and the plants are very invasive in Scotland. It is unusual, growing as a woody shrub rather than with canes. I would also avoid the many 'ground cover raspberries' promoted for landscaping and forest gardening. Their yields are poor, they can be invasive and they take up space better used by more productive plants.

Thimbleberries
Rubus, Subgenus *Anoplobatus*

ROSACEAE

I first came across thimbleberry doing forestry research fieldwork in the Upper Peninsula of Michigan. Back home, I managed to get some seed and have grown them; they grow vigorously but only fruit in particularly hot summers.

Japanese wineberry fruit with sticky wrapping

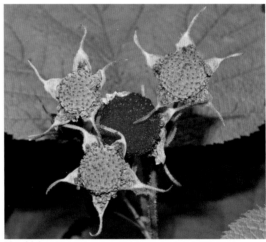
Thimbleberries

Learning a little more, I discovered that there is not one thimbleberry but two. The Upper Peninsula species, *R. parviflorus*, grows across the Great Lakes region and in all the western states of North America, including British Columbia and Alaska. Its East Coast cousin, *R. odoratus*, also known as purple-flowered raspberry and Virginia raspberry, grows from Alabama to Nova Scotia. It turns out that this is the plant I bought as 'thimbleberry'. The northern parts of both plants' ranges are well into the cool temperate zone, so northern strains should be expected to do well in other parts of the world. Thimbleberries have never been bred like blackberries and raspberries, probably because the fruit is too soft for commercial cultivation, but they deserve their foraging fans and a place in a home garden.

Hybrid berries
Rubus crosses

ROSACEAE

Berries cross not only within the subgenera (in which case they are usually marketed as cultivars rather than crosses) but between them too, giving rise to the hybrid berries. Crosses from Scotland have been given river names, such as the tayberry and tummelberry – both raspberry-blackberry crosses. The loganberry (*R.* x *loganobaccus*) is similarly a cross between a red raspberry and *R. ursinus*, a North American species of blackberry. Amateurs can play too: the hildaberry is a cross between a tayberry and a boysenberry (itself a raspberry-loganberry cross) discovered by a gardener in the 1980s. The list seems to be ever-growing, with many new breeds being crosses of crosses. I even have one hybrid berry in my own garden that just appeared – perhaps it's the alanberry!

By and large, hybrid berries grow more like brambles than raspberries. For the record, my favourite so far is the tummelberry.

Black chokeberry
Aronia melanocarpa

ROSACEAE

Chokeberries have beautiful flowers, striking autumn colours and edible fruit; all they need is a better marketing department! Actually, I'd add that the fruit itself could do with a bit of work too. It's high in pectin and useful for adding to other fruit for jams and jellies, but until it's fully ripe it's horribly astringent – and in my garden the birds strip them long before they are ripe. In my experience they are not worth planting unless you go for one of the more compact cultivars such as 'Morton' or 'Nero' (growing to about a metre) and net them in autumn. If you can get ripe fruit, you could try making wine with them, like the Orkney Wine Company, who use the appropriately-named 'Viking' cultivar. Just maybe don't call it chokeberry wine.

There is another species, *Aronia arbutifolia*, and a hybrid between the two, *A.* x *prunifolia*, but their fruit is not generally rated as highly as that of *melanocarpa*. All three make large shrubs of about 3m (9.8ft) unless you get a smaller cultivar.

Aronia melanocarpa berries

The Orkney Wine Company's rosé wine is made with aronia, cranberries, rosehips and rhubarb, all grown in Orkney

Flowering quinces
Chaenomeles species

ROSACEAE

Flowering quinces are hardy, but for fruit they need a warm, sunny position; this explains why they are often grown in Scotland as ornamentals but rarely for their fruit, which is similar to that of their close relative, *Cydonia*, and needs ripening inside. The Edible Leeds blog gives an enticing recipe for fermented quince pickle, a superior substitute for lime pickle. There are several species but *C. cathayensis*, the Chinese quince, is considered to have the best flavour. It makes a bush of about 3m (9.8ft).

Quince flowers

Roses
Rosa species

ROSACEAE

The dog rose (*Rosa canina*) is a Scottish forager's favourite, and I suspect that the wild roses are best kept that way. They will grow well and happily in a forest garden, but not as a well behaved member of the community. The growth potential of wild roses is phenomenal, leaving you with many metres of straggly, viciously-thorned briar to contend with every year. Some, like *Rosa rugosa*, are more modest in their top growth, but then they tend to spread aggressively at the roots.

If you do decide to put up with their habits, they will provide you with fragrant petals that can be made into rose petal jam or added to your bath, followed by their fruit: the rosehips. I mostly enjoy rosehips as a snack while walking. The irritating hairs mixed in with their seeds make the preparation of large amounts of hips tedious (and give them the Scots name of itchycoos). A way round this is to simply dry whole hips and use them to make tea. *Rosa rugosa* has some of the largest hips; it is widely planted as a municipal shrub so it is usually easy to forage.

Dried rosehips for making tea

Hoverfly getting stuck into a rose flower

Aralia family

Udo
Aralia cordata

ARALIACEAE

Udo is one of the most impressive sansai, growing from overwintering roots to a height of about 3m (9.8ft) every year. There are several edible parts: the roots, the young shoots, the pith of older (but still growing) shoots, the shoot tips and the flower shoots. All parts share a distinctive lemony-resiny flavour, which is strong enough to need tempering when prepared for eating. Shoot tips make great tempura and go well in stir-fries. The pith, with the bitterly resinous skin pared away, is usually soaked in water for a while to moderate the flavour, then sliced into salads, cooked with soy sauce and mirin to make kinpira or diced into a stir-fry. The flavour can also be moderated by blanching them in spring, and Stephen Barstow, in *Around the World in 80 Plants*, describes a minor industry sprouting udo roots in caverns under Tokyo.[1] The large roots, which can be dug in the winter, have a milder flavour than the top growth.

Udo is hardier than its Plants For A Future entry suggests, growing well for me in Aberdeen, Stephen Barstow in Malvik and also in Finland.[2] It is often sold in Europe as the cultivar 'Sun King' (a positive spin on its rather peely-wally yellow colour), which is less vigorous and somewhat less hardy. I prefer the wild type, but the cultivar might be better for a smaller space. There are several other closely related species. Manchurian spikenard (*A. continentalis*) is very similar and even hardier; my plant manages to set seed while my udo doesn't. I haven't been able to get hold of *miyamo udo*, the high mountain udo (*A. glabra*), but it also sounds promising. All udos tolerate a fair degree of shade, making them perfect for an awkward, shady corner in a garden, but they also revel in full sun.

American spikenard
Aralia racemosa

ARALIACEAE

American spikenard is similar to udo and can be used in all the same ways. It is not quite so big, but still large, growing to about 1.8m (5.9ft). Its root has a liquorice flavour (and has been used in making

Udo pith and shoots

My brother in front of an udo plant in the Cruickshank Botanic Garden, giving an idea of its size

root beer) and it is unusual in the family in that the fruit has been recorded as being used.[3] I haven't had a chance to test this myself as mine hasn't set fruit yet, but North American foraging sites mention them being used for jelly, wine and raw fruit and say that they taste like root beer.[4] It will grow in heavy shade if all you want is the roots and shoots, but it probably needs sun for fruit production.

Ukogi
Eleutherococcus sieboldianus and *pubescens*

ARALIACEAE

Ukogi, or five-leaved aralias, are spiny woodland-edge shrubs growing to around 3m (9.8ft). David Brussell reports the use of the shoots as tempura;[5] mine has shoots that seem rather small for this but taste nice and stir-fry well. As well as *E. sieboldianus*, which is fairly widely available in the West as a hedging and ornamental plant (including in variegated forms), Brussell lists *E. pubescens* as growing in the same prefecture and at the same elevation, so it should be hardy too.

Devil's club
Oplopanax horridus

ARALIACEAE

I would think very carefully before introducing the devil's club into a forest garden anywhere, especially outside its native range. It was and still is used widely in traditional medicine in the Pacific North West and nowadays it is sometimes marketed as 'Alaskan ginseng'. Plants For A Future says that young shoots are peeled and eaten, but only on the basis of some rather weak sources. The downside is that the plant is not named *horridus* for nothing. It is a large plant, covered in irritating spines, and it layers itself to form dense thickets. A closely related plant, *O. japonicus*, is found in Japan, where it does not appear to have been used as a sansai.

Ukogi leaves

Heath family

The heath family supplies a number of berry-bearing shrubs. All of them are acid-lovers that should never be limed. They generally have very small seed, sometimes like dust, which needs to be sown on the surface of an ericaceous compost and watered well.

Tall blueberries
Species in *Vaccinium*
Subgenus *Vaccinium*

ERICACEAE

Don't spend too much time learning the Latin names of the blueberries as the genus *Vaccinium* is ripe for rearrangement by botanists. The common names of the genus aren't much better, with a host of regional British names for blueberries (such as blaeberries, billberries, huckleberries and farkleberries) applied unsystematically to a range of North American species encountered by colonists. Fortunately a blueberry by any other name tastes as sweet, and we're more interested in the fruit than the names. Most blueberries are low-growing, but some deserve inclusion in the shrub category.

The best of the set is the highbush blueberry, *V. corymbosum*, which grows naturally in wet woods and pine barrens in eastern North America. It has become the basis of a major industry and widely hybridised, so that cultivars now have the best characteristics of several species. The berries are popular on cereal and yoghurt, in ice cream, in muffins and pies and as jam. They dry well and make an excellent fruit leather. The twigs and leaves (ideally with fruit attached) can also be dried and make a nice tea.

The biggest challenge is getting to the fruit before the birds. If they find your bushes you may need to net them, which is easy enough as they grow to 2m (6.5ft) at most, with most cultivars being smaller. Highbush blueberries grow in a wide range of habitats, from wooded to open and from wet to dry, so long as the soil isn't alkaline. In the forest garden they will take moderate shade next to a tree. I have found that they will fruit quite well even with tall herbaceous vegetation growing up around them, which makes for an alternative method of hiding them from the birds.

Other woodland species worth growing include *V. parvifolium*, the red huckleberry/bilberry, which grows to 1.8m (5.9ft) in coastal forests from Alaska to California, and is described as 'acid but very palatable'. *V. membranaceum*, the mountain or thinleaf huckleberry, grows in thickets and woodland edges in damp woods in Michigan and from Alaska to California. It reaches 1m (3.2ft) and the

Blueberries with raspberries and alpine strawberries

berries are among the largest and best-flavoured of all wild blueberries. *V. ovalifolium*, the black huckleberry, grows across northern North America in open woods and has a pleasant sweet flavour. It is taller than most, growing to 3m, and is particularly hardy.

More unusually, *V. ovatum*, the box blueberry, is an attractive evergreen shrub growing to 2.5m (8.2ft). Its berries are small, shiny, juicy and abundant but not very flavourful. The Caucasian whortleberry (*V. arctostaphylos*) comes from the Eastern Mediterranean and Western Asia. Its berries are also not as flavourful as highbush or thinleaf blueberries, but are juicy and refreshing, superior to the box blueberry. It grows to 3m (9.8ft) and has spectacular autumn colour.

Black huckleberry
Gaylussacia baccata

ERICACEAE

Gaylussacia is a genus of shrubs closely related to *Vaccinium*, and botanists may yet decide to merge them. The majority grow in tropical South America and the fruit of many is described as insipid, but one stands out for use in cool temperate forest gardens: *G. baccata*, the black huckleberry. It grows as far north as Newfoundland, is a forest plant especially associated with pine barrens (pine forests on dry, sandy soils) and has fruit liked by everyone. Sam Thayer describes it as "like blueberries, but better", which I didn't think was possible.[6]

Black huckleberry is easily propagated by division, but for some reason plants are very difficult to source in the UK. It is not easy to grow from seed; seeds are short-lived and need warm-cold-warm stratification, so try to obtain fresh seed in summer when they fruit. It spreads quite quickly at the roots and birds are bound to spread the seed, so it has the potential to become invasive in Scotland.

Salal
Gaultheria shallon

ERICACEAE

Salal is a low evergreen shrub growing on a spreading rhizome, sometimes planted as an ornamental in Scotland. The berries, which are produced in abundance over a long season as they ripen sequentially in the bunch, are like softer-textured, milder-flavoured blueberries. They mix well with other berries, go nicely on porridge and have been grown in Orkney for fruit wine. They dry well but make a very poor fruit leather. Picking them is a little fiddly as they don't come off the stem very easily and have a tendency to squidge. I find it

easiest to wait until a whole bunch ripens and pick it, stripping the fruit off later when I plan to use it. I'm not aware of any improved cultivars but have noticed quite a bit of variation in fruit size and flavour so there is definitely potential for improvement. It is another woodland edge species, tolerating shade but not fruiting so well in heavy shade.

Porridge mix: brambles and salal berries

Edible cake decorations

Prickly heath in flower

Prickly heath
Gaultheria mucronata

ERICACEAE

Prickly heath is a little bit of a novelty: it produces glossy, round, white-to-pink berries described by a friend as "edible cake decorations". They can be sweet and juicy if picked at the right time but have no strong flavour. They are widely planted as ornamentals so they make a good candidate for municipal foraging if you don't have space in your garden. It is a thick, evergreen shrub growing to 1.5m (4.9ft). There is a cross with salal called *G.* x *wisleyensis*, the fruit of which is more baroque looking but no more strongly flavoured.

Other families

Viburnums
Viburnum species

ADOXACEAE

The viburnums seem to specialise in colourful common names, including mooseberry, nannyberry, sheepberry, hobbleberry, possumhaw, withe-rod, arrowwood and wild raisin, testifying to a range of uses. The native Scottish species, guelder rose or *V. opulus*, is often included in lists of edible plants (and gets a taste rating of 3/5 on Plants For A Future), but I do not think it is worthwhile. The fruit is sour with an unpleasant aftertaste, and is more seed than fruit. It is a little better after bletting all winter (if the birds don't get the berries), but still not worth taking up space in a home garden for. I'm not the only one who thinks this way: Sam Thayer describes guelder rose simply as 'inedible'.[7] Unfortunately this experience, and the purchase of yet another *V. opulus* masquerading as the more palatable *V. trilobum*, put me off trying any further species in the genus, so the following is based on Thayer's account of North American species.

Two North American species both go by the name 'highbush cranberry'. They are no relation to the true (or 'lowbush') cranberry *Vaccinium macrocarpon*. *V. trilobum* is closely related to guelder rose and some biologists put them in the same species, which has in turn led to much confusion between the two species in the nursery trade. Wild *V. trilobum* vary considerably in quality, so if you want to grow it it's a good idea to seek out a superior cultivar like 'Wentworth', 'Andrews', or 'Hahs', to avoid both a poor-fruited form and the dreaded *opulus*. An even better bet is *V. edule*, also known as mooseberry or squashberry, which has a more northerly distribution and is reputed to be the better tasting of the two, but unfortunately seed is very difficult to source. I cannot find any record of any improved cultivars of this species.

The juice of highbush cranberries is refreshingly tart, but the seeds are unpleasantly bitter, so the juice must be extracted without tainting it with the bitterness of the seeds. This is done by gently mashing the raw, ripe fruit, then straining through a colander for pulp or a cloth for juice. These can be used to make jam, jelly or sauce. Juice can be clarified by leaving it to settle in the fridge for a few days. As it has natural antibiotic properties it will keep for a long time in the fridge.

An even more intriguing viburnum is *V. lentago*, known as nannyberry, wild raisin or black haw. It produces a mealy fruit, with a taste like a mix of banana and prune. The fruit, which is not ready until October, is separated from the seeds by simmering them for about an hour with a little water and then straining. The result is a thick purée which can be turned into fruit leather, used like banana in

Guelder rose in flower

Guelder rose fruit after winter

baking or used directly as a spread. Berries last a long time on the plant or in the fridge and can be frozen for later puréeing. The drawback to nannyberry is that its natural distribution only extends to the most southerly parts of Canada, so it may not be productive in Scottish gardens.

All the viburnums described here are large shrubs that grow best in damp soil. They are very ornamental, with attractive white flowers and red autumn colours, if somewhat straggly in form. They root easily by layering.

Saltbush

Saltbush leaves

Saltbush
Atriplex halimus

AMARANTHACEAE

I first came across saltbush at Plants For A Future's site in Cornwall. I loved the salty-tasting leaves, and spent many years trying and failing to grow it in Scotland. The tricky part is overwintering it, and the secret seems to be to plant in spring to give it plenty of time to get established and place it in a raised, sunny spot with very good drainage. In case of failure, you can take a backup copy by rooting a cutting, which is very easy, and keeping it inside over winter. The cultivar 'Cascais' has larger leaves and shorter internodes, making it ideal for production. Pick the shoot tips rather than just leaves, which also helps to keep the plant compact and stimulates new fresh growth. They go very well in a mixed salad. The bush eventually grows to about 2m (6.5ft).

Climbing spinach
Hablitzia tamnoides

AMARANTHACEAE

Also known as Caucasian spinach, climbing spinach is exactly what its name suggests: a tender leaf that grows as a climber. Although it comes from the Caucasus mountains, it has a history of being grown

Hablitzia tamnoides scrambling up the south-facing wall of a house in Norway. The purple flowers are everlasting pea (*Lathyrus latifolius*).

143

in Scandinavia, as told by Stephen Barstow in *Around the World in 80 Plants*.[8] Scandinavian-origin plants are probably the best bet for growing in cool temperate areas.

It is possible to grow climbing spinach up a tree, but it takes a lot of patience. It doesn't really like a lot of shade and it establishes much faster growing up a wall or supports in a sunny spot, in which situation it can easily grow to 3m (9.8ft). As well as the young leaves, which have a mild taste and can be harvested through much of the growing season, it produces a cluster of overwintering shoots at the base which can be harvested judiciously during the winter months.

Lovage
Levisticum officinale

APICAEAE

In terms of use, lovage belongs with the many celery-family members of the ground flora, but in terms of size it pushes its way into the middle layer.

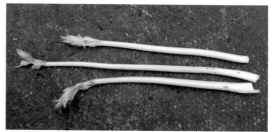
Blanched lovage shoots

It is a large herbaceous perennial best known as a herb useful for adding a savoury, meaty/yeasty taste rather like Maggi sauce to dishes, but growing a plant this size for the occasional fragment of leaf never made much sense to me. The more useful part is the spring shoots, which are more nutritious and milder. They can be chopped into sauces and stews or cooked as tempura and dipped in soy sauce. Blanching them by putting a bucket over the plant in early spring yields long, white shoots that are milder still. I also use it as a 'dual use' green manure plant, cutting the tall stems down for compost in summer.

Lovage shoots emerging

Darwin's barberry in flower

Barberries
Berberis species

BERBERIDACEAE

There are many barberry species, most of which
produce small, sour, orange fruits guarded by spines.
They have been touted as a lemon substitute but I
mostly find them too fiddly and too well-protected
to bother with (Martin Crawford calls them "sacrificial
species", a nice way of saying that the birds will
eat them all). The exception is Darwin's barberry,
B. darwinii, which fruits heavily, is relatively easy
to pick and makes nice jam. In contrast to most
barberries it has dark blue fruit and evergreen
foliage like tiny holly leaves, making it an attractive
shrubbery plant at all times of year.

Sweetshrub
Calycanthus floridus

CALYCANTHACEAE

Sweetshrub is well named. The bark and twigs
have a sweet smell reminiscent of cinnamon and
can be used as a substitute. The herbalist Juliet
Blankespoor uses them in tea and mead;[9] I use
them to flavour a daal. Be careful though: all other

Darwin's barberry fruit

parts of the plant, such as the leaves and flowers,
are poisonous and should not be used despite
their alluring scent. It overwinters without any
trouble in my garden. Although in its native habitat
it grows in riverine woodlands, it is best planted in
full sun here to help it ripen its wood and avoid frost
damage. It makes a sizeable shrub of up to 3m (9.8ft)
but can be trimmed smaller to save space and
harvest at the same time. *C. floridus*, the eastern
sweetshrub, has a western cousin, *C. occidentalis*,
but it is less hardy and not so easy to get hold of.

145

Hops
Humulus lupulus

CANNABACEAE

Hops may be best known for making beer, but they are also an excellent perennial vegetable. They have been described as the world's most expensive vegetable, apparently fetching up to 1,000 euros per kilo. I find this rather astonishing as they are actually quite easy to grow.

They are herbaceous climbers, dying down to the ground every year and climbing anew into the sun, and in the garden you need to provide a structure for them. Traditional hops grow to several metres so for an ordinary garden it is a good idea to get one of the new dwarf cultivars such as 'Prima Donna'.

The peak time for picking hop shoots is in late spring when the young shoots start to emerge from the ground. The plants spread by underground rhizomes, so this can sometimes be in unexpected places (see illustration, page 2). Harvesting is therefore combined with heading off a potential weed problem, cutting unwanted growing points down to the ground. I also find that the young growing tips are useable, if a little smaller and less productive, throughout the summer until the plant slows its growth, toughens up and turns its mind to flowering in the autumn. The constant nipping out of growing tips as you harvest them helps to keep the plant smaller, bushier and, by delaying flowering, tender for longer.

I usually harvest tips of about 10cm (4in) – longer than this and they are already becoming woody. The raw shoots are quite astringent so I always cook them, upon which they develop a lovely nutty flavour. In their early growth they produce enough to cook as a standalone vegetable. The cooking options are quite like green asparagus, which they are often compared to. They can be steamed or boiled, then served with butter or olive oil, or fried. They go very well blanched and then cooked in an omelette – if a little of the astringency is left for this dish it complements the egg well. Later in the year mine tend to go in stir-fries. Professional hop pickers can develop a sensitivity to handling the shoots but this is unlikely in quantities found in a home garden.

Honeyberry
Lonicera caerulea

CAPRIFOLIACEAE

As a large shrub with edible berries, the honeyberry is not what you expect of the honeysuckle family. It is very hardy and in recent years a wide

If you would really rather make beer

Honeyberry flower: time to get the paintbrush out

range of improved cultivars have become available, but new growers often find them disappointing. There are a number of reasons for this. Firstly, they grow slowly and take several years to come into production, so a little patience is needed. Then there is pollination. They are not self-fertile so two compatible cultivars are needed and since they aren't widespread you can't rely on plants in anyone else's garden. They seem to go out of their way to avoid pollination, producing rather inconspicuous flowers in late winter or very early spring; the RHS recommends hand pollinating them with a paintbrush! Some people also report that the fruit tastes like turpentine.

You can increase your odds by planting a number of different cultivars to increase the chance of pollination. Having other species of honeysuckle around might also help. If you have other early-flowering shrubs like willows around you'll attract more pollinators, who might then also notice your honeyberries. The turpentine taste is more of a mystery. Both growers who said this had their plants in rather shady places so it might be down to not ripening properly, or it might be a case of different people tasting the same fruit differently. My berries, when I can persuade the plants to produce them, taste pleasant, rather like blueberries.

Sea buckthorn
Hippophae rhamnoides

ELAEAGNACEAE

Sea buckthorn is an unusual plant in many ways. It is one of the few to fix its own nitrogen, a trait that helps it to thrive in a wide range of habitats and areas, from the Scottish coast to the Himalayas and the steppes of Central Asia (Genghis Khan was apparently a fan). It holds its bright orange fruit (which can crop exceptionally heavily in a good year) in an edible state for an unusually long time, from late summer through to the following spring. At the start of this period the little berries are moderately firm and can be picked, rather time-consumingly, in the normal way. As time

An unusually heavily yielding sea buckthorn bush from Scotland

goes on they become progressively more liquid and burst at a touch. This lends itself to an unusual harvesting method: a large vessel is held under a branch and the juice is simply squeezed from the berries directly into it. The branches are thorny so heavy gloves are recommended for this operation. A number of companies in Scotland, such as 'Wild and Scottish', sell wild-harvested juice from coastal stands.

The fruit is unusual in another way: it is both oily and juicy, giving it a silky texture and adding to its nutritional value. It is exceptionally high in vitamin C and is drunk by many people as a tonic (sometimes in the sense of 'gin and'). It is tart and zingy, too much to drink straight in volume. The chef Billy Boyter of The Cellar in Anstruther describes it as "Scotland's answer to passion fruit".[10]

Sea buckthorn is a large shrub, growing to 2-4m (6.5-13.1ft). In garden soils individual stems sometimes aren't especially stable and can fall over, but plants have a thick, vigorous root system that throws up new shoots. They are dioecious, requiring both a male plant for pollination and female ones for fruit. A wide range of cultivars have now been bred, mostly in Russia, Germany and the Baltic countries, for qualities like bush size, fruit size, yield and lack of thorns. The most widely available two in British nurseries currently seem to be 'Leikora' (f) and 'Pollmix' (m). Other female cultivars include 'Botanica', 'Garden's Gift' and 'Star of Altai'.

Soapberry
Shepherdia canadensis

ELAEAGNACEAE

Soapberry is a North American species closely related to sea buckthorn, sharing its silvery leaves, shrubby form, nitrogen-fixing abilities and unusual fruit. Its berries are rich in saponins, soapy chemicals which give a bitter taste and which mean that you shouldn't eat too many at once, but which are turned to advantage in the production of a dish known as Indian ice cream.

To make this, ripe berries are beaten together with sweeter fruit like juneberries or raspberries and (in modern times) a little sugar to raise a bittersweet froth. It is important not to get any oil in the mix as this will prevent frothing. Nancy Turner writes that, "The taste of Soapberries, like that of beer or pickles, is acquired: few people enjoy Indian ice-cream the first time they sample it. Even the sweetened froth has a sour-bitter taste, and an inexperienced eater is usually bothered by the air from the whip accumulating in the stomach. But once you overcome your initial dislike for it, you may find that it can be a novel and enjoyable treat." For those that acquired the taste, Indian ice cream was a sufficiently important treat to make preserved soapberries an important trade item for both coastal and interior nations in British Columbia.[11]

Soapberry is also known as russet buffaloberry. A relative, the silver buffaloberry (*S. argentea*), is less hardy and prefers a more continental climate. According to Daniel Moerman's *Native American Food Plants*, it was widely used as a dried fruit and to make jelly.[12]

Currants and gooseberries
Ribes species

GROSSULARIACEAE

Red-, white- and pinkcurrants (*Ribes rubrum*), blackcurrants (*R. nigrum*) and gooseberries (*R. uva-crispum*) are all natural woodland edge plants that grow well in the forest garden, fruiting well even in light shade. Improved cultivars of all of them have been bred, with larger, sweeter fruit and, in the case of gooseberries, fewer thorns. All benefit from pruning (see Chapter 4).

The birds are fond of their fruit, especially the redcurrants, so they need to be planted where they can be netted when the fruit ripens. This is particularly true since picking currants is fiddly, but easier if they can be left for the whole bunch to ripen than if you have to pick the ripe ones out from amongst the unripe ones. Some people put a sheet on the

Red, pink and white currants

Golden currant

ground and give the bush a sharp shake, but I find that the time saved in picking is lost in taking out all the leaves and insects that come down too. Yellow or 'white' varieties can also help reduce the amount lost to the birds.

There are a few pests and diseases to look out for. Blackcurrants often get 'big bud', in which a mite lays its eggs in the buds and causes them to deform from their usual pointy shape to a spherical form like a tiny cabbage. The mites also spread 'reversion virus', which causes leaves to be smaller and fewer-lobed and yields to be reduced drastically. They should be inspected every winter and the disease literally nipped in the bud. In badly

infested plants the cycle can sometimes be arrested by pruning the whole plant to the ground in the winter. If there are signs of reversion virus then the plant should be removed.

With gooseberries the biggest potential problem is the gooseberry sawfly, the larvae of which can completely strip the leaves of a plant. This won't kill the plant unless it happens year after year, but it will seriously reduce the yield. The larvae, which look like long, pale green, black-speckled caterpillars, can be picked off if damage is noticed. Gooseberries are also rather prone to mildew, which flourishes in damp, still air. You can find resistant cultivars or plant in an open position and prune to open up the bush and increase airflow. Infected shoots should be pruned off and disposed of.

All species in the genus are rich in pectin, which helps jams and jellies set, so they are often mixed with other fruit to help achieve this. Most of my redcurrants go into my raspberry jam. Gooseberries are a good carrier of other flavours; I make gooseberry and meadowsweet jam most years as they are available at the same time in the season.

All species have a tendency to sourness and it is difficult to eat lots of the raw fruit. Whitecurrants are the sweetest and I like to pick them by the bunch and put them in the fruit bowl. Eating a bunch at a time by stripping them through your teeth is very satisfying. Like this they last a good while; picking them off the bunch tears the skin and they rot much sooner. Blackcurrants make a nice fruit leather, especially mixed with a sweeter fruit like Japanese plums. Redcurrants and gooseberries have larger seeds that harden in drying and ruin the texture. Blackcurrants also freeze better than the others.

All species propagate easily from cuttings. Their lifestyle in the wild relies on this: they grow particularly in wet ground by forest streams and root into new sediments brought down by floods. It is a great way to acquire new cultivars, but be careful that you are not acquiring a disease at the same time.

There are many other species and a variety of crosses, although none of them compare with the traditional three in my opinion. One interesting option is *Ribes aureum*, the golden currant. It has attractive (and edible) yellow flowers and strong autumn colour, giving it more ornamental value than the others. Its fruit is larger than a black-currant and produced later than any of the other currants, extending the season into autumn.

Nalca
Gunnera tinctoria

GUNNERACEAE

Nalca looks like a monstrous rhubarb, growing as a herbaceous perennial to almost 2m (6.5ft) with leaves covered in warty spines. Its relative *Gunnera manicata* is often grown in gardens as a striking specimen plant. Nalca has similar cultural requirements, preferring a damp soil and a dry mulch over its crown over winter. It is easily propagated

Nalca patch and flower in Chile

by division and it can be worth keeping an offset indoors during winter in case the main plant is lost. It comes from Chile, where its petioles (leaf stems) are eaten by the Mapuche people – 'nalca' simply means 'petiole' in Mapuche. The petioles need to be peeled to remove the warty rind and can then be eaten raw or cooked. They have a mild flavour. On the West Coast and islands of Scotland, nalca has become invasive in the wild and is better foraged than cultivated.

Northern bayberry
Myrica pensylvanica

MYRICEAE

The northern bayberry is a large, nitrogen-fixing shrub growing in the north east of North America as far north as Newfoundland. I should warn you that it has one of the most confusing scientific names in the entire literature. It is *M. pensylvanica*, like Pennsylvania without the double 'n', but there is also a *M. pennsylvanica*, with the double 'n', which is a synonym for the much less hardy southern bayberry.

The useful part of the bayberry is the leaves, which are used as an alternative (foragers say a superior one) to bay leaves for flavouring food and drinks. Plants grown in full sun have more flavour than ones in shade. They are hardy plants, growing in a range of soils including wet ones and tolerant of salt spray. Historically the berries were used to yield wax for making candles, but since they are dioecious you'll need a male and a female plant if you really want to try this. The Scottish native *Myrica gale* (p.263) is a related, smaller option.

Fuchsia
Fuchsia species

ONAGRACEAE

Better known as ornamental shrubs with masses of flowers called *pendientes de la reina* (the queen's earrings) in Spanish, fuchsias also have nice fruit,

The development of fuchsia fruit. At the base of the flowers you can see small holes where short-tongued bumblebees have taken a shortcut to the nectar.

so long as you wait until they are fully ripe. Before this they taste horrible and will leave a prickly feeling in your mouth. *F. magellanica* is the hardiest – in fact it has become invasive on parts of the Scottish west coast. It is a large shrub which grows in shade but prefers a warm, sunny spot for fruit production. Its fruit doesn't change colour much as it ripens, but after a few unpleasant experiences with nearly-ripe fruit you soon get your eye in to the subtle changes in texture that announce ripeness. Fuchsias are easily propagated from cuttings, so if you find a superior variety do share it around.

Poke
Phytolacca americana

PHYTOLACCACEAE

Poke, also known as pokeweed, has a challenging reputation as one of the tastiest wild vegetables in North America – and also one of the most

poisonous if not prepared correctly. It is one of the most well-known wild edibles in the US and until 2000 there was even a minor industry canning wild-collected poke. It is a large herbaceous perennial, growing to over 2m (6.5ft). Sam Thayer describes finding plants that are decades old,[13] but my experience of it in Aberdeen, both in my own garden and in the Cruickshank Botanic Garden, is that it is a short-lived perennial, maintaining itself by seeding but not living for very long as an individual plant.

The safe and worthwhile part of a poke plant is the vigorously-growing, tender, meristematic stem shoot, which should bend and snap easily. In spring the entire shoot can be harvested, being careful not to include any of the root. Later in the year, while the stems are still growing vigorously, the lower parts will have hardened off and become unsafe to eat but the growing tips are still useable. Foraging books often specify a length of stem that

Poke flowers are very pretty, but the shoots are not edible at this stage

Poke fruit is so black that it has been used to make ink!

is safe to pick, but this varies with the size of the plant. Thayer says that there may be 50cm (20in) of tender shoot on a large plant.[14] By contrast the useable length on a small plant may be shorter than any specified amount. The important thing is to understand how to identify meristematic tissue (Chapter 1).

A poke stem is really two vegetables: the leafy tips, which are used as a pot herb, and the tender stems, which can be peeled and used as a shoot vegetable. Both parts must be detoxified by cooking twice. They should first be boiled in plenty of water for 10 minutes. Thick stems should be peeled and cut into chunks for this, to ensure that all parts are well leached. The poke should then be drained well and the cooking water disposed of immediately, as people have become sick from drinking it. The shoots or greens should then be cooked again. For added safety, they can be parboiled a second time, in a second batch of water, cooking for two minutes. This is definitely recommended for the first time that you prepare poke. After this, the leaves can be used as a pot herb and the stems as a vegetable in a variety of ways. The traditional Appalachian method is to fry parboiled shoots with bacon and onions, stirring in and cooking an egg at the end.

Poke is sometimes confusingly called poke salad or sallet. This preserves the old meaning of the word, signifying a vegetable. *No part of poke should ever be eaten raw.*

The colour of the skin on poke stems varies from red to green. Foraging lore holds that red stems indicate higher levels of toxins and it is often recommended that plants with any red in their stems should be avoided. Thayer points out that "very young shoots of perfect quality, when growing in full sun, typically have a reddish hue."[15] If the shoots are reddish he either peels them or boils in two changes of water. If you are cultivating rather than foraging poke, it would be worth selecting and breeding from plants with the greenest stems.

Bamboos
Subfamily Bambusoideae

POACEAE

In general grasses are best kept out of the forest garden, but one subfamily, the bamboos, are different. The young shoots are so high in protein that pandas have the same protein intake as carnivorous bears, despite their vegetarian diet. For bamboo shoots, canes should be cut at a height of about 18cm (6in), cutting 5cm (2in) below the surface to get the maximum length of tender shoot. The shoot is sliced down the middle and the tender core can be lifted out of its sheath of leaves.

Bamboo shoots

Bamboo shoots cut

All bamboo shoots are edible if prepared right, but, as usual, the question is not 'is it edible?' but 'is it worth it?' Many bamboos contain cyanogenic glycosides, toxic chemicals which must be broken down by cooking. Some contain very little or none and can be eaten raw; some contain a bit more and can be made palatable by boiling for a short while; yet others contain lots and need boiling in multiple changes of water. A shoot that is horribly bitter when raw can be delicious once cooked. Size also matters; some shoots are just too small and fiddly to be worth processing.

Bamboos also vary considerably in how they behave in the garden. Some grow in a tight clump that only slowly expands; others run, producing new shoots at up to a metre (3.3ft) away from the last ones. Bamboo rhizome is very tough and hard to dig out, so you don't want to make a mistake and have to remove a bamboo from half your garden. Rhizome barriers dug into the ground can help to contain aggressive species, but it is much easier not to plant them in the first place. Spreading behaviour depends partly on climate: *Phyllostachys* species have a reputation for running in warmer parts of the world but are reasonably well behaved in cooler climates. *Sasa kurilensis*, although considered to be a good edible bamboo, is extremely invasive.

Besides the edible shoots, bamboos obviously yield mature canes, which are used in China for everything from building to weaving. They're pretty useful in the garden too. As mature plants, the canes make great supports and burn like wax when dry.

The table overleaf shows a range of edible bamboos, with some judgements on their edibility. It is evident that opinions vary, which may be influenced by personal taste but probably also depends on whether the comment applies to the raw or the cooked shoot. Shoots that are described as bitter will require longer boiling than others. In the table below, Lewis Bamboo refers to the taste ratings given on the website of the US bamboo nursery of that name.[16] Guadua Bamboo cultivate and export giant timber bamboos in Latin America,[17] and Martin Crawford's comments are from his book *Creating a Forest Garden*. 'PFAF' is Plants For A Future, while 'SB' refers to Scottish Bamboo, a bamboo nursery based near Turriff in Aberdeenshire. They tell me that all the bamboos they offer are proven to grow well on their site, which is a good reality check for whether a species will flourish in cool temperate conditions. Unfortunately, two of the most prized bamboos in Asia, *P. dulcis* (sweetshoot bamboo) and *P. edulis* (moso) fail the Turriff Test and did not survive in my more coastal garden in Aberdeen either.

Species English name Chinese name	PFAF hardiness rating	SB	Uses mentioned in the E-flora of China	PFAF edibility rating	Lewis Bamboo	Guadua Bamboo	Martin Crawford
Chimonobambusa marmorea kan-chiku 寒竹 han zhu	5-9	Y	no uses mentioned	1		delicious	
Chimonobambusa quadrangularis square bamboo 方竹 fang zhu	5-9	Y	no uses mentioned	2		delicious	
Chimonobambusa tumidissinoda walking stick bamboo 筇竹 qiong zhu		Y	edible shoots			delicious	
Phyllostachys atrovaginata incense bamboo 乌芽竹 wu ya zhu		Y	edible, weaving			edible	
Phyllostachys aurea fishpole bamboo, golden bamboo 人面竹 ren mian zhu	6-11	Y	ornamental, no mention of shoots	5	**		free of bitterness – good
Phyllostachys aureosulcata yellow-groove bamboo 黄槽竹 huang cao zhu	5-9	Y	ornamental, no mention of shoots	4	**		free of bitterness
Phyllostachys bambusoides madake 桂竹 gui zhu	6-9	Y	As *P. reticulata*: building, bitter shoots	4	**	bitter (Note 3)	bitter when raw
Phyllostachys bissetii 蓉城竹 rong cheng zhu	4-8	N	tool handles, weaving, very hardy	0	**	edible	
Phyllostachys flexuosa zigzag bamboo 曲竿竹 qu gan zhu	5-9	N	delicious	3		delicious	very bitter raw
Phyllostachys glauca blue bamboo 淡竹 dan zhu		Y	weaving, furniture and shelters	3		good	slight bitterness
Phyllostachys heteroclada water bamboo 水竹 shui zhu		Y	widely cultivated for weaving		**	edible	
Phyllostachys iridescens red chicken bamboo 红哺鸡竹 hong bu ji zhu		Y	delicious	3		delicious	
Phyllostachys mannii (Note 1) beautiful bamboo 美竹 mei zhu		Y	weaving		**	edible	
Phyllostachys nidularia bignode bamboo 篌竹 hou zhu	6-9	N	shoots edible, ornamental	5		delicious	slight bitterness

Species English name Chinese name	PFAF hardiness rating	SB	Uses mentioned in the E-flora of China	PFAF edibility rating	Lewis Bamboo	Guadua Bamboo	Martin Crawford
Phyllostachys nigra black bamboo 紫竹 zi zhu	7-10	Y	widely cultivated, no uses mentioned	4	*****	delicious	bitter when raw
Phyllostachys nuda grey bamboo 灰竹 hui zhu	7-10	Y	delicious, very hardy	4	*****	delicious	v slight bitterness – good
Phyllostachys propinqua morning garden bamboo 早园竹 zao yuan zhu	7-10	Y	hardy, weaving, no mention of shoots	3		good (Note 4)	
Phyllostachys prominens giant timber bamboo 高节竹 gao jie zhu		Y	shoots delicious			good	
Phyllostachys rubromarginata reddish bamboo 红边竹 hong bian zhu	7-10	Y	cultivated in Henan, no uses mentioned	3	**	edible	good quality
Phyllostachys violascens (Note 2) violet bamboo 早竹 zao zhu	6-10	N	early spring shoots	3		delicious	slight bitterness
Phyllostachys viridiglaucescens greenwax golden bamboo 粉绿竹 fen lü zhu	6-9	N	shoots delicious	4		edible	free of bitterness – good
Phyllostachys vivax giant timber bamboo 乌哺鸡竹 wu bu ji zhu	7-10	Y	widely cultivated for shoots	3		delicious	slightly bitter
Pleioblastus simonii Simon bamboo, medake	5-9	Y		3			fair quality
Pseudosasa japonica arrow bamboo, metake 矢竹 shi zhu	5-9	Y	cultivated as ornamental	2			free of bitterness – good
Semiarundinaria fatsuosa Narihira bamboo 业平竹 ye ping zhu	6-9	Y	native to Japan, no uses mentioned	5			slightly bitter
Yushania anceps Indian fountain bamboo	8-11	Y		0			free of bitterness – good
Yushania maling malingo	8-11	N		0			good

Notes
1. Listed on Scottish Bamboos and Guadua Bamboo as *P. decora*.
2. Listed on Plants For A Future as *P. praecox*.
3. The form 'Shouzhu' is described as 'edible'.
4. The form 'Lagusinosa' is described as 'delicious'.

Giant knotweeds
Reynoutria species

POLYGONACEAE

I'm definitely not suggesting that you plant Japanese knotweed (*R. japonicus*) in a forest garden in Scotland. Its lack of respect for the built structures of humans and habit of simply growing through, rather than round, them, means that the cut plant and soil that may contain the rhizomes are classified as controlled waste, and you can even be on the receiving end of an Anti-social Behaviour Order for growing it. I'd just like to mention that in its native range in Japan it is a prized sansai, with shoots that taste rather like rhubarb. The less known Sakhalin knotweed (*R. sachalinensis*) is similar.

Greenbriers
Smilax species

SMILACACEAE

Smilax species are near the top of my list of plants that I want to get hold of but haven't managed to so far. Sarsaparilla, the scent of the American South and of many a short story by Ray Bradbury, belongs in this genus, but only a few species are hardy in cool temperate areas. These include the China root (*S. china*) in East Asia, and the cat greenbriar (*S. glauca*), carrion flower (*S. herbacea*), hag brier (*S. hispida*), horse brier (*S. rotundifolia*) and Blue Ridge carrion flower (*S. lasioneura*) in North America. All of these except *lasioneura* have been recorded as having edible roots, but the prize part is the spring shoots, which are large and taste somewhat like asparagus. Sam Thayer recommends the shoots of the carrion flower very highly and

A present from a friend. It was delicious

suggests that we look past its unappealing common name and the revolting smell of its flowers (designed to attract pollinating flies, not us).

Greenbriers are not tidy plants. Their growth habit is to grow up and then fall onto surrounding vegetation, which they either cling to with tendrils or simply lie on for support. Given the promise of their shoots, I'll put up with this if I can get one.

Goji berry
Lycium barbarum

SOLANCEAE

Ignore the marketing: goji berry is not worth growing in a cool temperate garden, although it will happily do so. It is a sprawling shrub with a tendency to spread by layering itself. I have yet to meet anyone in the north of Scotland who has managed to get berries from it. The shoots are edible but not very exciting.

Bladdernuts
Staphylea pinnata and *trifolia*

STAPHYLEACEAE

The bladdernuts (one in Europe and one in North America) are large woodland shrubs that bear small nuts inside a papery bladder. They are fiddly to crack; Martin Crawford puts them through a nut-cracking machine on a small setting.[18] *S. trifolia*, the American bladdernut, is the better bet for a cool temperate garden, with a natural range extending to southern Ontario and Quebec.

In a survey of wild food plants used in two valleys of the Qinling Mountains in central China,[19] the related species *S. bumalda* and *holocarpa*, both of which are sold in the UK as ornamental shrubs, were amongst the most common species named by villagers. The leaves are used and can be preserved by drying.

Smilax herbacea shoot

Wild garlic flower

9

Crops for Forest Shade

When we get down to the low-growing crops below the trees and shrubs, the degree of shade becomes all-important. This is the most diverse layer of the forest garden, so I'll divide the plants according to their shade requirements. This chapter deals with the crops that can tolerate shade for most of the day. Some of these are true woodland species that do best in the shade. Others might grow faster with more sun but are best in a shady part of the forest garden to make them more tender or restrain their spreading tendencies.

Good King Henry
Chenopodium bonus-henricus

AMARANTHACEAE

As a herbaceous perennial member of the spinach family, Good King Henry (GKH to its friends) makes a predictable appearance in books and seed catalogues featuring perennial vegetables. Many of these almost guarantee to put the reader off the plant by saying that the young leaves are edible in salads. They are not; they are covered in bitter-tasting saponins and I have never met anyone who likes them. Those who do use the leaves prefer them cooked and tend to be further south; I speculate that the further south you go and the faster it grows, the milder-tasting it gets. There also seems to be an element of individual sensitivity to saponins. I know one couple who respectively love and hate leaves from exactly the same plant. Personally, I find that there are so many nicer-tasting

leaves available by the time GKH sprouts that I haven't the slightest inclination to use it.

It can still be worth growing, however, for two other products. The young flower shoots are relatively chunky and easy to peel. Without the soapy skin they are very mild flavoured and have a look that explains their nickname of 'Lincolnshire asparagus'. But perhaps the most interesting aspect of GKH is the potential of its seeds as a temperate, perennial version of quinoa (*Chenopodium quinoa*). Tycho Holcomb of Myrrhis Permaculture[1] harvests about a kilo of seed each year from his forest garden, cleaning the saponins off the seeds by running them several times through a hand blender with water. His online video takes you through all the steps from harvesting to eating. Grow in shade for leaves and shoots, in lighter shade for seeds.

Good King Henry

'Lincolnshire asparagus' on forest garden risotto

Snowbell
Allium triquetrum

AMARYLLIDACEAE

Years ago, walking on a Cornish coastal path, I found some bulbs that had been dug up, perhaps by a badger, and stuck them in my pocket. For some time they grew in my garden with the label 'Cornish mystery allium'. I now know that they are snowbells, also known as three-cornered leeks. They do look rather like a white bluebell and it is important

Snowbells and (poisonous) bluebells growing together in the Cruickshank Botanical Garden. Not a good idea in a forest garden.

not to confuse the leaves: snowbells have a different pattern of flower attachment, onion-scented leaves and a green stripe on each petal.

One useful aspect of snowbell is its growth pattern. It starts into growth in autumn, will grow right through a mild winter when few other alliums are available, and dies down in early summer. The leaves can be used like leeks or onion tops, the flowers look nice in a salad and the bulbs have a mild garlic flavour. It is carpet-forming, spreading by division and seeding, but in the north of Scotland it is not so invasive as it can be further south. The few-flowered leek, *A. paradoxum*, is similar but more invasive and more fiddly to harvest; I would not recommend it.

Wild garlic / ramps
Allium ursinum, tricoccum

AMARYLLIDACEAE

Allium ursinum and *tricoccum* are not especially closely related within the alliums, but they share an ecological niche and a common name in Europe and North America respectively. The European species, known as wild garlic, bear garlic, ramsons or ramps, is an allium with broad, soft leaves, adapted to growing in heavy shade in moist woodland soils. All parts are edible and have a strong garlic flavour. It is among the first of the herbaceous perennials to start growing again in spring, reappearing in my garden in late February or early March. I love its leaves chopped into salads or layered into a sandwich. Wild garlic pesto packs quite a punch and it is far better than bulb garlic for hummus (which you can make truly local by using broad beans instead of chickpeas). You can make a superior (and far easier) garlic bread simply by sprinkling finely chopped wild garlic onto a piece of toast and drizzling it with olive oil.

For all this, what makes wild garlic really useful is that when cooked it loses its garlicky flavour and becomes instead a mild, oniony pot herb. This turns it from a herb to a bulk vegetable and makes

A serious patch of wild garlic

Wild garlic clumps

In the frying pan

Ramps

Wild garlic dies back dramatically at the first touch of hot sun, so it is worth having some planted in a really shady spot, which will extend the season to around June. Once it dies back you can also harvest the bulbs, although they are on the fiddly side and not the best garlic bulb available. They do not store like the bulbs of cultivated garlic as they dry out easily and die quite quickly if they are not stored moist. They can be preserved by pickling.

When Scots and Irish settlers went to North America, they applied their local name – ramps – for wild garlic to a similar-looking woodland allium, which is also known as wild leek. It has become a strong part of Appalachian culture, where ramps festivals are part of the calendar in many areas. Unlike wild garlic, its leaves become fibrous and unuseable once flowering starts, but its bulbs are shallower-growing and better. Both species make a great sauerkraut-like ferment.

Ground elder
Aegopodium podagraria

APIACEAE

Ground elder might normally be thought of as a weed that gardeners spend a lot of time trying to remove from their gardens, but it is also a tasty perennial vegetable, as might be expected by something brought to Britain by Italians (going by the name of Romans back then). Its common Scottish name of bishop's weed also hints at its culinary past as it was grown in the gardens of monasteries and bishops' palaces.

If you want to try growing it in a forest garden, don't forget its spready nature. I have contained it in one bed for a decade by using all the tricks. It is in full shade, which reduces its energy for spreading and makes the leaves more tender. It is a variegated variety, which again is less vigorous and therefore less invasive, but no less tasty. It is hemmed in by hoed woodchip paths and interplanted with other 'thugs' like wild garlic and wild strawberry. Even with these precautions, it is

its productive, carpeting growth a virtue rather than a problem. In spring I use it wherever I would use onion. The only drawback is that if it dries out while cooking, it quickly acquires an unpleasant burnt-onion flavour.

You can harvest wild garlic simply by pulling off individual leaves or, for less garlicky hands and to speed things up, you can cut a clump at a time with scissors. I generally put my wild garlic leaves in a bowl of cold water for five minutes as soon as I get home, to preserve and wash them. They'll then keep for at least a week in the fridge. Another way of harvesting that gives a slightly different product is to dig up a clump and then prepare the individual plants by cutting off the roots and removing the sheath of the bulb. The whole thing then hangs together in a sort of 'spring onion' version of wild garlic. Fried in plenty of oil and dipped in sauce, these are gourmet food indeed.

Ground elder

Ground elder leaves ready for picking

important to prevent it from ever flowering and seeding; fortunately the young flower stems are delicious, so removing them is its own reward.

As we'll see again and again, especially with the celery family, the edible part is the young, emergent leaf shoots, picked before they have fully expanded. They are best fried in olive oil, and every spring I make a dish that I call 'pernicious pasta', in honour of both ground elder's Roman connection and its reputation as a pernicious weed. I break a handful of dried linguine in half, leaving it around the same length as the ground elder shoots, boil until nearly al dente, drain and put aside. I then fry a small onion (chopped) for a couple of minutes, add a few chopped mushrooms and fry for two minutes more. To this I add some nettle tops and fry for five minutes more, then I add ground elder leaf shoots and fry for another five minutes until tender. I then add the linguine and stir, followed by some cream and a little of the cooking water from the pasta, a teaspoon of bouillon or other stock powder, and fresh, finely-chopped herbs such as parsley, wild celery, wild garlic, Scots lovage and sweet cicely. After a couple more minutes' cooking it is ready to serve.

Ground elder flowers cooked as namul

Pernicious pasta

Years of eating it have given me a different perspective on ground elder's weedy nature: it's an edible plant that is very productive, grows strongly enough to outcompete any weeds and tolerates shade and poor soils. What's not to like?

Pignuts
Conopodium majus and *Bunium bulbocastanum*

APIACEAE

I got very enthusiastic about pignut (*Conopodium majus*) in my early days of forest gardening, even naming my blog about it: 'Of Plums and Pignuts'. They are shade-loving plants that produce a starchy edible corm that I find tastes like hazelnuts (although some people detect a less pleasant hint of radish). I have always enjoyed foraging them and looked forward to growing extra-large ones in my allotment. Unfortunately it turns out that pignuts are excellently adapted to growing with limited resources and, given optimal conditions, continue with their frugal ways unaltered. A seed needs winter stratification to germinate, after which it unfurls its seed-leaves, forms a tiny corm and dies down until next year. The corm doesn't divide like bulbs or tubers, just grows slightly larger every year, making propagation difficult. It buries itself deep, coming to the surface by wandering, thread-like stems, so it can be hard to harvest. I now think of pignuts as an occasional treat that will maintain itself in your garden if allowed to self-seed rather than as a significant crop.

There is also a semi-mythical plant called greater pignut (*Bunium bulbocastanum*), which grows wild around Cambridge in England. In the British nursery trade it seems that plants sold as greater pignut are almost always *Oenanthe pimpinelloides*, the corky-fruited water-dropwort. I have known a number of people who have eaten these and speak well of them, but care should be taken as there are some very poisonous plants in the same genus.

A pignut with spring growth from the corm

Pignuts peeled and sliced

Sweet cicely
Myrrhis odorata

APIACEAE

Sweet cicely is best known for its long, aniseed-flavoured seeds, which make a nice wayside nibble on a summer's day, but it has the potential to be far more productive than this. Another traditional use is to add the leaves to rhubarb jam; they contain a natural sugar-free sweetener which will take the sourness off the rhubarb as well as adding that distinctive flavour. My main use for it, however, is the young leaf-shoots which, as usual, are milder and tenderer than the mature leaves. I put them in soups, stews, salads, stir-fries and tempura. The immature flower stems are even better than the leaf shoots, and more solid. Plants grown in full sun tend to stop production of new shoots in summer, putting out a new flush in autumn, but ones in shade will produce new shoots all season long. Since it is one of the first plants to sprout after winter and one of the last to die down, you will only be without shoots for the hardest months of winter.

At this point you can turn to the roots, which contain a particularly strong dose of the aniseed flavour. This can be used almost as a spice to flavour winter stews, or chunks of the root can be fried long enough to drive off most of the aniseed flavour and leave a starchy chip. Mature sweet cicely roots are massive. I once tried to dig up one whole: I got down to 1.2m (4ft) before it snapped off, at which point it was still about 2.5cm (1in) thick. One-year-old roots are more manageable and better textured so it can be worth sowing some and treating them as an annual crop. If you sow too many, the seedlings are also tasty cooked whole. Sweet cicely has a strong requirement for winter stratification, so think ahead and sow in the autumn.

My main complaint about sweet cicely is that the local (but not native) strain found in Scotland has hairy leaves with an unpleasant mouth-feel. Luckily, it turns out that a Danish strain is perfectly hairless. I was given some seeds by Søren Holt,

Sweet cicely in a shady corner of the forest garden

Sweet cicely seeds

Leaf shoots for cooking

165

Honewort

Mitsuba 'atropurpurea'

who writes the blog 'In the Toad's Garden'.[2] They grow just as well in Scotland and seem to have straighter roots than the locals too.

Honewort and mitsuba
Cryptotaenia canadensis and *japonica*

APIACEAE

Honewort and mitsuba are two closely-related woodland plants from North America and Asia respectively. In Japan, mitsuba began as a sansai but it is now widely cultivated, and used in a range of dishes from soups and salads to tempura and sushi. I find the taste of both plants somewhat reminiscent of ground elder and use the young leaf stems and flower shoots in similar ways. In North America honewort has a reputation for making great stock for soups and broths. The forager Steve Brill dries it for use in bean soups and Sam Thayer says that "few other herbs produce such a rich, complete-tasting broth all by themselves".[3] The roots are small but delicious.

Stephen Barstow finds the purple form of mitsuba, 'Atropurpurea', hardy in his Norwegian garden down to about -15°C (5°F).[4] Honewort is even hardier and both succeed in my garden in Aberdeen. Perhaps the hardest part of growing

cryptotaenias is getting them started. Like many Apiaceae, the seeds need winter cold and don't store well, so try to get fresh seed in autumn and sow immediately. They are perennial so once established they will last. They will also self-seed and can even become invasive, so a wise precaution is to eat the flower shoots before they can seed (although if you do let it seed, the seedlings are also edible). They grow to about 30cm (12in) and purple mitsuba is an attractive ornamental as well as a food plant.

Mayflowers and Solomon's seals
Maianthemum and *Polygonatum* species

ASPARAGACEAE

Mayflowers and Solomon's seals are related to the lilies of the valley, which are similar-looking but definitely not edible. The mayflowers are also sometimes called false Solomon's seals, false lily of the valley and false spikenard. It must be hard to be so regularly accused of being an imposter. Besides being fairer in every sense, 'mayflower' is closer to the Latin name.

Two mayflowers stand out for edibility: the feathery mayflower or treacleberry (*M. racemosum*) and the starry mayflower (*M. stellatum*). Edible

Mayflower flowers, in May

Mayflower shoots: feathery mayflower (top) and (starry) mayflower bottom

Solomon's seals include *P. biflorum*, *commutatum*, *multiflorum*, *odoratum*, and *verticillatum*. *P. commutatum* is now regarded as a large variety of *P. biflorum*, but it is still worth trying to get hold of for its size.

Both groups throw up annual shoots from a rhizome. These rhizomes have also been reported as edible but they are slow growing and bitter tasting and require a lot of processing: famine food only! By contrast the young shoots are sweet and edible with a slight bitterness that some enjoy and some don't (I do). The leaves are very bitter and should be removed; in the case of Solomon's seal I remove the whole stem above the first leaf pair. Pick while leaves are still rolled up and the shoot pointing upwards. Later on, the leaves unfurl and the shoot arches over: Solomon's seal flowers hang beneath while mayflowers have frothy clusters of blossom at the shoot tips. At this stage they are no longer worth eating. This makes for a short picking season, which I sometimes miss.

I eat the stems fried or boiled. The fruit of *M. racemosum* is said by some sources to have a delicious, bittersweet flavour reminiscent of treacle. Sam Thayer finds the aftertaste bitter and acrid.[5] It appears that this varies between local populations of the plant: Nancy Turner reports that the berries are eaten by some peoples in British Columbia but

Solomon's seals look like curious snakes when they first emerge

Bumblebee on Solomon's seal flowers

167

disdained by others. They are called 'sugarberries' in Carrier territory and used to sweeten other foods.[6] Mine flower but have never fruited, so I don't know where I stand on the issue yet.

Waterleafs
Hydrophyllum species

BORAGINACEAE

The eastern waterleaf, *H. virginianum*, is a welcome sight as it unfurls its leaves in early spring. It is a mild-tasting salad plant, with a slight bitterness in a similar sort of way to lettuce. Its name comes

Hydrophyllum virginiana leaf with watermark

The young flower shoots of Jack-by-the-hedge are the nicest part

from the 'watermark' blotches on its leaves, which make it attractive as well as tasty in the salad bowl. Mine has been quite well behaved, but like many in its family it has a reputation for being invasive, so make sure that you can confine it and be careful about letting it seed. Older leaves, flower stems and buds are all nice cooked, but later in the year they become too bitter. Besides the eastern waterleaf, there are *H. canadense*, *capitatum* and *appendiculatum*. The last of these is biennial; Stephen Barstow finds that it maintains itself by seeding in his garden.[7]

Jack-by-the-hedge
Alliaria petiolata

BRASSICACEAE

The flavour of Jack-by-the-hedge is best described by its alternative name: garlic mustard. It is a Eurasian native which can be somewhat spready in a forest garden, so only grow it if you are keen on the taste. In North America, don't touch it with a bargepole: it is a serious invasive pest that sets a nasty ecological trap for butterflies which lay their eggs on it thinking that it is the native toothwort (*Dentaria*). When the eggs hatch the caterpillars are unable to digest the leaves and they soon die. Its roots have a nice, horseradish-like flavour but are quite fiddly to harvest.

Giant bellflower
Campanula latifolia

CAMPANULACEAE

The bellflowers are a large genus adapted to many niches. One of the best is a true woodland plant: *C. latifolia*, the giant bellflower. It is an untidy plant, shooting up to a height of two metres from perennial roots and then flopping all over the place, so staking is a good idea. The growing tips are, as usual, the best bit, with a sweet, nutty flavour either raw or cooked. The roots are sweet and crisp and the leaves and flowers can both be used in a salad. Stephen Barstow blanches the shoots, which makes them even sweeter and more tender.[8] He describes giant bellflower as being "both … cabbage and carrots" to the Sámi people of northern Scandinavia.[9] It can spread by both seed and root to become invasive, but this isn't generally a problem if you are eating them. There are more bellflowers in Chapter 10.

Bunchberry
Cornus canadensis

CORNACEAE

Bunchberry is also known as creeping dogwood. Its beautiful flowers, with four porcelain-white bracts, reveal its relationship to the shrubs and trees of the dogwood family, but it instead makes a low-growing herbaceous perennial, spreading by slender rhizomes. It is very hardy, growing even in Greenland. It grows well in my garden but unfortunately it is self-sterile, so I'll need to get another, genetically distinct, plant before I can try the bunches of berries that give it its name. These are generally agreed to be not the most exciting fruit in the world, but pleasant and sweet. They are also high in pectin, so they make a useful base for helping other fruit to set.

Cairmeal
Lathyrus linifolius

FABACEAE

Cairmeal tubers (pronounced like 'Carmel') have an interesting reputation as an appetite suppressant used by Highlanders to keep hunger pangs at bay during a long journey or working day. To a forest gardener, the most interesting thing about cairmeal is its combination of nitrogen-fixing powers, shade tolerance and edible tuber. Although it grows well enough in my garden I haven't succeeded in getting worthwhile tubers, but Gregory Kenicer writes in *A Handbook of Scotland's Wild Harvests* that "bitter vetch [cairmeal] is a great candidate for controlled

Giant bellflower

Bunchberry in flower

Cairmeal flowers

domestication, as seeds germinate readily and it can be grown in a large pot or in a wild bed or woodland corner of a garden."[10]

The tubers are 0.5-2cm (0.1-0.7in) in diameter and have a flavour somewhere between fresh peas and mild liquorice. Besides eating fresh, they have been used to flavour alcohol. The young shoots can also be used, like little pea shoots.

Lemon balm
Melissa officinalis

LAMIACEAE

Lemon balm is certainly the easiest way to grow lemon flavour in cool temperate gardens: much simpler than lemon trees or lemongrass! It is a bomb-proof herbaceous perennial whose main drawback is that it can do too well for the amount of it that you need. I relegate mine to deep shade, which it tolerates perfectly well, to slow it down. The flavour is heat stable and can be used in cooking wherever you want a lemon taste. The fresh shoots make particularly good tempura.

Mints
Mentha species

LAMIACEAE

Just like lemon balm, mints provide a strong and distinctive flavour for teas, salads and cooking, but can overstay their welcome as they spread strongly through masses of underground rhizomes. Some forest gardeners grow enormous quantities of mint to attract pollinators and distract pests, but I think that any well-functioning forest garden should do this automatically. I like to keep my mint proportional to the amount I will actually use, which in practice means restricting it to large pots, kept in shady parts of the garden.

Mints come in many different varieties and flavours and it is difficult not to get collector's mania. I restrict myself now to spearmint (*M. spicata*), peppermint (*M.* x *piperata*) and apple mint

Lemon balm

Garden mint

(*M.* x *villosa alopecuroides*), all of which have different qualities of minty flavour. Oh, and I couldn't resist chocolate mint, a cultivar of peppermint. I learned from a herbalist friend that mint tea with a blend of varieties is much nicer than with any one alone.

Peppermint being restrained from taking over the whole bed

All kinds of pollinators love mint flowers

Some other members of the same family can also be used for a mint flavour. Ground ivy (*Glechoma hederacea*) has a flavour that is harsh and unpleasant on its own but serves to round out the flavour of mint tea superbly. It is a native woodland perennial with strong spreading tendencies.

The mountain mints (*Picnanthemum*) have a lovely flavour and are less invasive than mint, but have never flourished in my garden. They are winterhardy but seem to need relatively hot spring temperatures to bring them back from hibernation; mine rarely came up before midsummer.

Dog's tooth violet in flower

Erythronium 'Pagoda'

Bulbs from 'Pagoda'

Fawn lilies
Erythronium species

LILIACEAE

The erythroniums fill an important niche in the temperate forest garden, as the only shade-tolerant group with sizeable starchy roots. They come in many forms. The European dog's tooth violet (*E. dens-canis*) is oddly named: it looks nothing like the true violets and isn't related to them, but the root does indeed look like a canine tooth. In Japan the ka-takuri (*E. japonicum*) is used as a source of starch for sauces, while *E. sibiricum* has been used in southern Siberia.[11] In North America the erythroniums are known as fawn or trout lilies, presumably from the mottled pattern on the leaves. Species include *E. albidum, americanum, grandiflorum, montanum* and *revoltum*, with *revoltum* and *grandiflorum* having the largest bulbs. According to Nancy Turner, Pacific Northwest First Peoples used to leave yellow avalanche lily (*grandiflorum*) bulbs to soften for two days and then dried them on strings; circular strings of dried bulbs were an important trading item.[12] For garden use, the best bet is perhaps the sterile hybrid 'Pagoda', which has larger bulbs than any of the species.

The bulbs don't have an outer skin like an onion does, which makes them easy to prepare but rather prone to drying out if you aren't careful with storage. They can be harvested at different times. The plants die down in June or July, leaving just enough top growth to locate the bulbs. At this stage they are moist and starchy. My favourite way of cooking them is to slice them thinly across and fry the discs. They go chewy and sweet, a bit like plantain chips. Another way of frying them is to make chips (in the British sense). The smaller bulbs are just the right size already; the larger ones can be sliced in half or quarter. They are also good boiled and excellent in stews. It's not something I've tried myself, but according to Plants For A Future, the European dog's tooth is dried to make flour and used in making cakes and pasta. An alternative is to harvest the bulbs in early spring, when the leaves first begin to show. At this point much of the starch has been mobilised into sugar, so they are far sweeter.

Erythroniums' straight-up-and-down growth habit means that they mix well with more sprawling plants like wild strawberries, and their early growth leaves half the growing season for another crop. Like most bulbs, they are adept at punching up through a thick layer of mulch, so I give mine a mulch of leaves in the autumn to both feed and protect them.

Scootberries
Streptopus species

LILIACEAE

The scootberries, also known as mandarins or twisted-stalks, are not easy plants to get established. The seeds are slow to germinate and then slow to grow. Division of the rhizome works well when they are small but divisions of older roots are apt to sprout and then die suddenly. It's worth persisting, however, for the novelty of a shade-loving herbaceous perennial that tastes in all its parts like cucumber or watermelon. The young shoots can be used either raw or cooked. The fruit should only be eaten sparingly though. Sylvanus Hayward wrote in 1891 that "*Streptopus roseus* [now *lanceolatus*]

I learned to call Scootberry long before I understood why it was so called. The sweetish berries were quite eagerly eaten by boys, always acting as a physic, and as the diarrhea was locally called 'the scoots' the plant at once received the name."[13] White scootberry berries can be cathartic, which also doesn't stop children from foraging them.[14] Species include the white scootberry (*S. amplexifolius*), pink scootberry (*S. lanceolatus*) and small scootberry (*S. streptopoides*). They are quite ornamental plants so a number of nurseries listed on the RHS Plant Finder sell them.

Spring beauties
Claytonia species

MONTIACEAE

Spring beauties are low-growing, spreading plants mostly native to North America and Asia. Most are perennials. *C. sibirica*, the Siberian spring beauty, better known as pink purslane, is widely naturalised in Scotland, where it self-seeds freely and forms carpets in shady woods. As a spring ephemeral it dominates the woods for a few weeks, then largely disappears for the rest of the year. There is a white form sometimes called the Stewarton flower as it is dominant around Stewarton in Ayrshire, but it is found in many other places too. The young

Pink scootberry

Pink purslane

leaves are eaten raw or cooked. In salads they have a flavour of raw beetroot that I don't like much, but some people do. Other species recorded as having edible leaves include *Claytonia acutifolia, caroliniana, exigua, lanceolata, megarhiza, scammaniana, tuberosa, umbellata* and *virginica*.

All of these except *exigua* and *scammania* have also been recorded as having little edible tubers,[15] described as tasting like radish raw but potato or chestnut cooked. The forager Euell Gibbons wrote of fairy spud (*C. virginica*) that "We tried them fried, mashed, in salads, and cooked with peas, like new potatoes. All these ways were completely successful, but, as regular fare, we preferred them just boiled 'in the jackets.' My friend grew so fond of this food that he was afraid he would experience withdrawal symptoms when the supply was exhausted."[16]

There is also an annual species, called miners' lettuce (*C. perfoliata*). Its leaves are mild-flavoured

Miners' lettuce's unusual flower head

and succulent so they make an excellent bulk ingredient for salads. All parts are edible, including the leaves, stems and the unusual-looking large fleshy bract around the flowers. They can also be cooked, for instance in stir-fries. It is rich in vitamin C;

Miners' lettuce seems to grow larger in a tangle with other plants than on its own

the name comes from its use against scurvy by gold miners in California's gold rush. There are two closely related species: *C. parviflora* and the deep red *C. rubra*. I can't find any information on the edibility of these but I'm sure they would be worth investigating; *C. rubra* in particular would look very striking in a salad.

Miners' lettuce is often grown as a greenhouse crop in Scotland but it will thrive outdoors too. I spent a long time trying to establish it in my garden from bought seed before discovering that it grew as a weed nearby. This local strain seems to do a lot better than the others in my garden so getting well-adapted plants for your area might be the key. They can germinate at various times, sometimes coming up in autumn and standing through the winter, sometimes waiting for spring. It will grow in the open or in partial shade and likes a well-watered soil. As an annual, it needs a degree of soil disturbance to maintain itself in an area.

Ferns
various species

CLASS: POLYPODIOPSIDA

Ferns are the dinosaurs of the plant world. Once dominant, they are now reduced in stature and extent, but within their new niche they still flourish with style and grace. They are now mostly ground layer plants in damp, shady places, which makes them perfect for the forest garden, although only a handful are edible. The useful part of most edible ferns is the 'fiddlehead', the emerging frond, while the end of it is still rolled up like the scrollwork on a fiddle. People sometimes imagine that it is only the rolled-up section that is edible, but it is the whole frond at this stage. Just eating the tips would be a great waste. Many ferns contain thiaminase, an enzyme that breaks down the vitamin thiamine and can cause deficiency if ingested in large quantities over a long period of time, but it is harmless

Ostrich fern (*Mattheuccia struthiopteris*) spreading by rhizome

in small quantities and destroyed by cooking and thorough drying.

Ferns have a complicated lifecycle, alternating between a large, asexual stage which produces spores, and a small, sexual stage which emerges from the spores and produces sperm and eggs to give birth to a new asexual stage. For this reason they are normally propagated by the simpler method of dividing the rhizome or root mass.

Bracken (*Pteridium aquilinum* and related species) is one of the most popular ferns for eating, but it would be unwise to let it into your garden, with its aggressively spreading rhizomes and habit of taking over entire hillsides. It is viewed with suspicion and little eaten in Britain due to the presence of a carcinogenic toxin, but in East Asian countries including China, Japan and Korea, the fiddleheads are eaten widely, either cooked fresh or preserved by salting, pickling, or sun drying. The toxic chemical, ptaquiloside, is water soluble and broken down by both heat and alkaline conditions, but suspicion remains that bracken might be implicated in Japan's relatively high incidence of stomach cancer. I have tried Scottish bracken fiddleheads and find them enjoyable, but I'll keep my consumption of them low. Starch from bracken rhizomes has been used in many cultures but Gordon Hillman was unable to extract useable quantities from ones harvested in Britain.[17]

My favourite edible fern for the forest garden is *Mattheuccia struthiopteris*, the ostrich fern. They are native to northern Asia, north and central Europe and northern North America, so it's something of an oddity that they aren't found naturally in Scotland. The fiddleheads are nice raw, steamed or fried. It's a bit of a cliché in perennial vegetable books that shoot vegetables are always described as tasting like asparagus, but ostrich fern really does. Like bracken, it spreads by rhizomes to form carpets but it is so edible that this is not a problem. They are plants of river-bottom woods and really don't like drying out, so give them a particularly moist, shady area or be prepared to water them in

Ostrich fiddleheads at the perfect point for picking

Sterile and fertile (centre) fronds on ostrich fern

Ostrich fern hit by dry conditions in the Royal Botanic Garden Edinburgh

long spells of dry weather. One advantage to ostrich fern is that it is a popular ornamental and plants are relatively easy to buy. There is a cultivar called 'Jumbo' which has larger fronds and emerges later than normal. In ideal conditions even regular ostrich ferns can grow up to 1.5m (4.9ft), although in the less-than-ideal conditions of my sandy soils they are more like a third of this size.

Another large, edible fern is *Osmunda japonica*, the Japanese royal fern. It is collected wild in China, Tibet, Japan and Korea and known as zenmai in Japanese and gobi in Korean. Like ostrich fern it is a plant of wet woodlands and it can only tolerate open sun in very wet soils. It grows successfully, but slowly, in my garden. It produces both fertile and sterile fronds, and it is only the fertile ones that are eaten.[18] The sterile fronds are large and spreading and have a tough texture, while the fertile ones are smaller and upright and more tender. Later on they will produce fuzzy brown 'sori' (spore production patches), while the fertile ones won't. These should help you distinguish the two types although of course neither is edible at this point.

Zenmai's North American relatives, *O. cinnamomea* (cinnamon fern) and *O. claytonia* (interrupted fern), are often described as edible but perhaps we should pay attention to Sam Thayer's account of eating them.

"Faced with so many people who claimed that these ferns were perfectly fine to eat, I decided to test my hypothesis [that the bitterness of both species indicated harmful compounds and that regular confusion between these ferns and ostrich ferns accounted for occasional reports of poisoning from the latter]. A friend and I consumed 10-12 raw interrupted fern fiddleheads one day in May. Not surprisingly, we got sick. Our symptoms were identical: severe headache, nausea, dizziness, lethargy, and general malaise. Later, I forced myself to eat a serving of cinnamon fern fiddleheads boiled. The symptoms were similar but milder, and the fiddleheads were so repulsive that I had to force myself to swallow them."[19]

Zenmai, showing mature fertile and infertile fronds

This might explain why there are no records of use of the closely related *O. regalis*, royal fern, which is native to Britain. Various other ferns have been recorded as edible but they are mostly considerably smaller, such as hard fern (*Blechnum spicant*), which is often found on rocks and walls in Scottish woods. Sweetfern (*Comptonia peregrine*), a North American tea plant, looks for all the world like a fern but is actually in the Myricaceae family, related to bog myrtle and bayberries.

Lesser celandine
Ficaria verna

RANUNCULACEAE

Around the world, members of the buttercup family herald the coming of new growth and are eaten as early greens and spring tonics. In Italy the shoots of the climber *Clematis vitalba*, known as vitalbini, are harvested and cooked into omelettes. In Japan the delicate woodland wildflower *Anemone flaccida*, called nirinsou, is collected as a sansai. In North America the abundant leaves of marsh marigold (*Caltha palustris*) are foraged as a potherb,[20] and columbines (*Aquilegia* species, Chapter 10) are widely used for both leaves and flowers.

It must be said, however, that the popularity of the family owes more to earliness than strictly to taste.

Anemone flaccida

Lesser celandine

Purple form of lesser celandine

It's a family that invests heavily in producing toxins and even the more edible members contain a chemical called protoanemonin that causes a violent burning sensation in the mouth and would make you quite sick if, against all sense and the firm advice coming from your mouth and throat, you managed to swallow a large quantity of it. Luckily, protoanemonin is destroyed by cooking or drying, but in some cases the amount of cooking required for taste reasons (boiling in multiple changes of water is recommended for marsh marigold) leaves something to be desired in terms of texture.

My favourite in the family is lesser celandine (*Ficaria verna*), a small plant with glossy leaves and cheery yellow flowers. It could be described as a buttercup that has reinvented itself as a woodland bulb. It bears fleshy little tubers that give it the resources for an early start into growth and underlie its invasiveness. They also explain lesser celandine's other common name: pilewort. Their shape was considered to resemble that of haemorrhoids or piles. Under the ancient 'doctrine of signatures', God was held to have marked each species to indicate its use to humans, so this resemblance was considered a sure-fire sign that celandine would cure piles. Miles Irving, the author of *The Forager's Handbook*, says that the tubers have a flavour and texture similar to potatoes and can be used boiled or roasted,[21] but they are so small and fiddly that this is one harvest that I'm prepared to leave in the ground.

The tubers also help lesser celandine to a well-deserved reputation for being invasive in damp or shady areas, where it can form extensive carpets. In North America, where it is introduced and where several states list it as a noxious invasive species, the cons definitely outweigh the pros. In European countries, where it is either native or a long-established introduction, the situation is different. In spite of its reputation, there are reasons why lesser celandine finds it difficult to become a serious pest in any well-managed garden. Despite the seeming ability of the tubers to get everywhere, it doesn't actually 'run', either underground like couch grass or overground like its cousin, creeping buttercup. It's also a very low growing plant. Its ambition is not to get into the full sun, so it rarely provides serious competition for other plants and it is really quite easy to weed out. It also has an Achilles' heel, which is that it needs constant moisture to stop the tubers drying out, and it's never going to be a problem in drier, sunnier areas of the garden.

The earliest lesser celandine leaves are very low in protoanemonin and are mild enough to use in a salad. As levels increase they become better cooked. I use them as a pot herb, in stir-fries (where they keep their succulent texture) or fried in olive oil until crispy. Traditional lore has it that you shouldn't eat lesser celandine after it has flowered, but I think that your taste buds are a better guide to protoanemonin levels than the flowers.

Some variations on the regular lesser celandine are available. There are varieties that do not produce tubers and are therefore much easier to control. I'm not sure, however, how easy this strain is to get hold of and whether or not it will tend to revert to tuberising as it self-seeds – I suspect so. There is also a handsome bronze variety which looks very striking with the bright yellow flowers against dark purple leaves.

Wild strawberries
Fragaria species

ROSACEAE

It's hard to know which section to put the strawberries in. They grow happily in full sun but will also bear heavy shade. I use wild strawberries to fill in what might otherwise be unproductive areas in shady beds. They spread by runners and can be as invasive as any weed in a fertile, sunny garden, but this can also be an advantage if you have a space to fill. Make sure, however, that when you plant a wild strawberry you know where you want it to stop. The fruits are no comparison in size or productivity

A strawberry starting to form in the centre of a flower

The final result

for the cultivated strawberry but win hands down on flavour. They rarely yield enough for bulk uses like jam, but are best picked a handful at a time to go on anything from porridge to cheesecake. If you have enough, they dry easily just by being left on a sunny windowsill and they impart a lovely flavour to fruit leathers.

The European wild strawberry (*F. vesca*) is the Scottish native. The green strawberry (*F. viridis*) and hautbois strawberry (*F. moschata*), both of which have hints of pineapple and musk in their flavour, grow in much of mainland Europe and Russia but are not native to the British Isles. Asia has a whole host, including the Japanese wild

strawberry (*N. nipponensis*). Eastern North America has the Virginia strawberry (*F. virginiana*), while the beach strawberry (*F. chiloensis*) grows on the Pacific coasts of both North and South America. These last two are particularly notable as the ancestors of the cultivated strawberry. This, and the alpine strawberry, are described in Chapter 10. Besides these there are some improved varieties of the wild strawberries: most usefully some yellow-fruited forms that the birds leave alone.

Nettles and wood nettles
Urtica and *Laportea* species

URTICACEAE

Nettle (*Urtica dioica*) is the Jekyll and Hyde of the home garden. It has no neutral qualities, only excellent and abominable ones. On the negative side, they spread aggressively by seed and underground runners, and attack anyone who dares try to weed them out with hypodermic syringes full of irritants. And as the leaves mature, they form microscopic stones called cystoliths that lodge in the kidney to form kidney stones. On the positive side, it is nutritious and tasty, yields fibre and medicine, dries and stores easily, is an important plant for wildlife and accumulates a suite of minerals that makes it one of the best plants around to go in your compost heap.

There is a long tradition of using nettles in Scotland. Bronze Age bog bodies have been recovered wearing clothes made from nettle fibre. St Colmcille, who spread Irish Christianity to Scotland in the sixth century, is reputed to have lived on nettle pottage after learning the recipe from an old woman he encountered cutting the plants. Legend tells that a servant, perhaps with less faith in the wisdom of old women, mixed meat juice into the broth from a hollow spoon that he used to stir it.[22] Samuel Pepys was served nettle porridge on his travels through the Highlands and Sir Walter Scott mentions the practice of forcing nettles under glass in *Rob Roy*. Anywhere in the world where nettles grow (and that is to say, almost every-where), there is a tradition of using them. St Colmcille might be given a run for his money as the patron saint of nettle eating by the Tibetan poet-sage Milarepa, who subsisted on them so much that his skin is said to have turned green!

Nettles have a distinctive, earthy taste that I didn't like when I first tried them but have grown to love. The part to pick is the tips: roughly 10cm (4in) of growth where they are still soft and break off easily. Wearing gloves helps if you don't want to be stung. You'll be glad to hear that even very brief cooking melts the stinging hairs and renders them powerless. My main use for nettle tops is as a pot herb: they add a depth of flavour to leaf sauce that I look forward to every year. Whole, they are good fried or steamed, either on their own or mixed with other spring shoots, or on top of a pizza. They make a great filling for ravioli or maki. The traditional nettle soup is always worth making, although I tend to round it out with other leaves. Make a potato soup, throw in nettle tops and cook for just a couple of minutes before blending. Miles Irving recommends instantly chilling the soup in a metal container plunged into cold water to preserve the colour and flavour.[23]

If you have more nettles than you can use they are easy to dry by spreading them out in a well-ventilated space. The dried tops can be crumbled and stored, to add nettly goodness to soups and stews throughout the year. Dried nettles also make a great tea, and blend well with mint. Fresh ones can be fermented into a refreshing 'beer', with no more added ingredients needed than sugar, yeast and a bit of lemon juice.[24]

Nettles are so common that it may be better to forage them than give up space in your garden for them, but there is one big advantage to having your own patch. As shoots grow older they develop harmful cystoliths (hard deposits in the cell walls) and they should not be picked after they begin flowering. You can extend the season by cutting the patch down. It won't be long before it is producing

tender regrowth. This makes nettle one of the best 'dual use' food/green-manure crops.

Alternatively, you might want to allow the plants to flower in order to try another product: nettle seeds. For culinary use, harvest the seed heads

Stinging nettles

Milarepa. I haven't experienced this side-effect of a nettle-rich diet so far.

while still green, then dry and rub the heads to release the seeds. They have a nutty, nettly taste and
can substitute for poppy seeds. Dry frying them
briefly before use brings out the flavour and melts
any stray stinging hairs that remain. If you feel
good after a spoonful, it might be because they are
packed with neurotransmitters such as acetylcho-
line and serotonin. The nettle isn't producing these
complex chemicals for our benefit but to increase
the potency of its sting, but we can turn the tables
by using its seed. This neuroactive quality means
that nettle seed consumption should not be over-
done. Forager and herbalist Monica Wilde writes
that a cup of nettle seed tea (boiled fresh nettle
seed in a 1:12 ratio, 25g (0.8oz) to 300ml (half a pint
water) may keep you wide awake for several days!
She recommends that you don't consume more
than 30g (1oz) per day in any form.[25]

Every area has its own nettle. Some local varieties
are recognised as separate species while others
are now regarded as subspecies or just local strains
of *Urtica dioica*. By and large it seems best to rely
on your local nettles and your local foraging
traditions. There are only two that might be worth
introducing if you don't have them locally. The first
is the fen nettle (*U. galeopsifolia*), which grows in
wet places across Europe. It is every forager's
dream: a stingless stinging nettle! Unfortunately it
crosses with stinging nettle and my plants at least
aren't completely sting-free, but they are much
less aggressive than the plants I am used to, and
some back-crossing might restore a pure stingless
population.

Another nettle you might want to try to source
is *Laportea canadensis*, the wood nettle. This close
relative is similar in many ways to the stinging
nettle, but according to Sam Thayer it tastes even
better, being "notably reminiscent of asparagus in
both flavor and texture". Anyone who has tried to
grow asparagus in cold areas will leap at the idea
of a plant that tastes like asparagus but grows like
a nettle. Wood nettle has a much fatter stem than
stinging nettles, so young shoots can be treated

as stem vegetables, either peeled or rubbed clean
to remove the stings (which, like those of *Urtica*,
melt in heat in any case). The shoots are tender
and juicy and solid all the way through, with a
mild, sweet flavour. Later on, the tops and then the
seeds can be used in all the same ways as *Urtica*
nettles.[26] Wood nettle grows as far north as Nova
Scotia and Stephen Barstow has succeeded in
cultivating it in his garden in Norway.[27]

Finally, Canadian clearweed (*Pilea pumila*) is a
member of the nettle family that has the virtue of
having no stings at all, and can even be used in
salads. It is collected as a sansai in Japan,[28] where it
is known as *aomizu*, but it doesn't seem to be valued much in the rest of its range, which includes
virtually all of eastern North America and East Asia.

All of these species are hard to get hold of
outside their native range. For some inexplicable
reason the nursery trade doesn't tend to deal in
nettles. One exception to this is the golden-leaved
dioica cultivar 'Good As Gold'.

All nettles will happily grow in full sunshine, but
in shade they grow larger, more tender leaves and
do not spread so aggressively. Sam Thayer reports
that wood nettles in deep shade tend to be smaller
than those in more open areas.

Violets
Viola species

VIOLACEAE

There are between 500 and 600 violet species
around the world, most of them in the temperate
north, but their range includes Australasia, Hawai'i,
the Andes and Patagonia. Many of them are eaten,
both for their aromatic flowers and for their leaves
and shoots, and some are shade-tolerant woodland perennials. In my garden I have settled on
V. riviniana, the native Scottish wood violet, of
which I have an attractive purple form. The leaves
and flowers are both attractive additions to salads,
although the leaves have a slight scratchy texture
that puts some people off them and they are quite

Low-growing wood violet under hedge woundwort (foreground) and *Tradescantia* (background)

time-consuming to pick. The young shoot tips are better, with a nutty flavour and a tendency for the plant to produce more of them when they are picked. It stands over winter but doesn't put on much growth at that point so pickings from it are limited. It is a low-growing, slowly-spreading plant that grows well with more upright plants.

The sweet violet (*V. odorata*) has larger leaves and gets the best taste score from Ken Fern,[29] but it is not very hardy. It is not considered native north of Westmorland and Durham and has died out several times over winter in my garden. However, the Online Atlas of the British and Irish flora has a number of records north of the Great Glen and I know of a patch growing wild in a sheltered spot on the Black Isle, so it may be a case of getting the right variety.

Wood violet leaves

Serbian bellflower

10

Forest Edge Crops

The next group of crops falls somewhere between the inhabitants of the deep forest and the sun-lovers of the next chapter. They are plants of forest glades and gaps created by fallen trees, of the forest edge or of scrub and open forest. Sometimes their preferred habitat might be open sun but the gardener gets better results, in the form of more productive, more tender plants, where they have to cope with some shade. They are the largest group of all, which is one of the reasons why we design to create plenty of suitable space for them in a home garden.

Sea beet and perpetual spinach
Beta vulgaris

AMARANTHACEAE

Sea beet is the wild ancestor of a host of modern crops, such as beetroot, sugar beet, fodder beets, spinach beet and Swiss chard. While its natural habitat might be the seashore, it fits perfectly into the forest edge niche of a home garden, with the light shade stimulating it to greater leaf production. It is a biennial (or sometimes a short-lived perennial)

'Sea beet' (left) and 'perpetual spinach' (right) leaves

Wild sea beet growing on the harbour wall on Eilean Eisdeal / Easdale Island

Sea beet flower shoots

but easily maintains itself by self-seeding if you let it. My garden population is a mongrel mix of wild sea beet, leafy spinach beet (or 'perpetual spinach') and the occasional chard that has been adapting to my site for 20 years and has grown noticeably more cold-tolerant in the time, growing right through recent winters (although the winters are themselves changing). Wild ancestry gives itself away in slightly smaller, waxier, darker green leaves compared to perpetual spinach.

The leaves of sea beet are an excellent pot herb, with a soft texture and rich flavour. They are produced throughout the first year of growth and a mild

winter. In spring it seems to take an odd pause just as other plants are getting going again, then starts again with renewed leaf growth followed soon after by chunky flower stems. At this point I switch my attention to the flower stems which, picked as early as possible, are excellent cooked Korean *namul* style.

Victory onions
Allium victorialis and *ochotense*

AMARYLLIDACEAE

These two closely related plants, one from Europe and one from East Asia, are similar to the

Victory onions emerging from the soil

A victory onion shoot

Victory onion flower

North American ramps (Chapter 9), but less shade-dependent. They have broad leaves – not quite so soft as wild garlic – and a long, thick, juicy stem. Make sure that you harvest the stem, not just the leaves, as the stem dies off and goes to waste if left. They can be used anywhere you want an oniony flavour and are particularly good for fermenting. A cultivar called 'Cantabria' has larger leaves and a more upright form.

Leeks
Allium porrum

AMARYLLIDACEAE

The leek is the queen of the Scottish vegetable garden in winter, standing through the cold months like nothing else. You can grow traditional biennial varieties in gaps in the forest garden, transplanted from seed trays started off inside in late winter, but there are also perennial cultivars which will come back year after year. I grow both, because the perennials are most productive in spring, just once you've dug up the last of the biennials. They produce a clump of familiar-looking leek stems, which can be cut an inch or two below ground level to get the maximum length of shoot while leaving the bulb in the ground.

There are some well-known lines of perennial leek, like the Babington leek, which grows wild on seashores in the south of England. Elephant garlic is another, despite its name and the mild garlic flavour of its large, juicy bulbs. However, most perennial leeks are clones, reproducing only by bulbils borne in the place of flowers. This leads to a narrower genetic base than I am comfortable with. Fortunately, perenniality is never far from the surface even in ordinary leeks. Leave any crop of leeks in the ground for a second year and a proportion of

A strongly bulbing perennial leek

Parts of a perennial leek for cooking

187

them will reshoot. This gives the opportunity to create landraces of seed-grown perennial leeks for different parts of the country. I have the beginnings of one in my garden, grown mostly from perennial leeks that popped up from my favourite traditional cultivar, 'Musselburgh'.

Growing my own leeks has restored my liking for them as a vegetable. Shop-bought leeks have often undergone a mass cleaning process and are saturated with fine grit. Fresh from the garden they are clean through. The whole shoot can be used, not just the white section. The scape, the young flower shoot, can also be used while it is young and tender.

Division from a non-bulbing perennial leek derived from 'Musselburgh'

A Babington leek flower head, with bulbils and flowers

Leeks have an unusually broad range of reproductive strategies. Underground they produce two kinds of bulbs: the normal ones like giant garlic cloves and smaller, harder outlying ones that may last in the soil for years before sprouting. Above ground they can choose between seed and two kinds of bulbils. The first, produced by strains such as 'Babington', are like tiny onion bulbs. The second are like tiny leeks and begin to grow in the flower head; they are known as 'grass'. Competitive leek growers (a big thing in Scotland and the north of England, really) sometimes 'shave' the flower heads of champion varieties, cutting off all the flowers to promote the production of grass as an emergency response. If you have a favourite variety, you will never be short of options for propagating it.

Jimbur
Allium wallichii

AMARYLLIDACEAE

A final allium for a forest edge niche is jimbur, from the Himalayas. It's unusual for an allium in that the roots are eaten,[1] although they are rather thin and have a stringy core so they aren't top of my list. The tops make a nice, oniony green, and they are available later in the year than wild garlic, snowbell or victory onion.

Jimbur and roots

Angelicas and alexanders
Angelica and *Smyrnium* species

APIACEAE

Garden angelica (*A. archangelica*) is a large herbaceous plant, technically a biennial but I often find that plants take a few years to get to seeding size and then will live for several years more if a lot of their flower shoots are harvested. The young leaf shoots, which are available in winter and spring, can be used sparingly as a distinctive flavouring for soups, stews or leaf sauce, but the best part is the peeled flower shoots which are sweet and juicy. Traditionally they were candied (see Chapter 5) and used in sweets, ice cream and cakes. In *Around the World in 80 Plants*, Stephen Barstow describes an old Norwegian cultivar called 'Vossakvann', which has solid stems and has been selected for sweetness, so that the leaf shoots can be used like flower shoots. They can be made even milder by blanching.[2]

Many other angelicas are edible too, including Korean angelica (*A. gigas*) and purplestem angelica (*A. atropurpurea*), both of which are striking plants with deep purple colouring. The roots of Chinese angelica (*A. sinensis*) are a popular spice[3] and bupin[4] in China. The Scottish native wild angelica (*A. sylvestris*) is a much smaller plant that I find incredibly bitter in all parts, although some people use the flower stems.

Angelica flowers are adored by pollinators, which practically queue up to wallow in the large blooms. The seeds need winter cold and don't last long in storage, but they have been coming up in my garden for years from a seeding almost a decade ago so they obviously last better in the ground.

Alexanders (*S. olusatrum*) are similar in their growth habits and can be used in all the same ways. They grow wild on the coasts of southern England but are surprisingly frost and shade-tolerant when they have to be. A patch in the Cruickshank Botanic Garden in Aberdeen thrives in heavy shade under a large sycamore tree and a wall. 'Alexanders' is

Angelica (left) and alexanders (right) flower stems

Peeled angelica flower stem

Purple angelica flower, Royal Botanic Garden Edinburgh

Alexanders root and shoot   Alexanders seeds

short for 'parsley of Alexandria', a name that goes back to the Romans. Its scent is supposed to be like that of myrrh – as in gold, frankincense and myrrh. This means nothing to me but was confirmed by my Austrian flatmate, since myrrh is still used there in Christmas celebrations. I now think of myrrh as 'the resin that smells like alexanders'.

Two additional uses of alexanders are the roots and the seeds. They have compact, chocolate-coloured roots rather like carrots. They are too strong-tasting to eat as a vegetable on their own but can be chopped into dishes for flavour. The seeds look like fat, black peppercorns and can be put into a pepper-grinder to add the aromatic taste of Alexandria over the top of any dish you like.

Those seeds are probably easier to grind than they are to germinate. They need winter stratification, preferably for two seasons. Mark your sowing row well and grow something else in it in the first year! The plants are smaller than angelica: around a foot or so, not counting the flowering stems. A second species, perforate alexanders (*S. perfoliatum*), is supposed to be milder-tasting than *olusatrum*. It's on my Christmas list.

Celeries
Apium species

APIACEAE

The ancestral wild celery, also called smallage or *Apium graveolens*, grows across the temperate northern hemisphere in damp, salty environments. It is used as a herb in some countries' cuisines, but it has also been bred into a number of different forms. These include var. *dulce*, the familiar shop-bought celery with its large, crunchy stems, and var. *rapaceum*, the turnip-like celeriac, with its swollen stem base. Of most interest in the forest garden, however, is var. *secalinum*, variously known as leaf, herb, cutting or Chinese celery. *Secalinums* are much closer to the wild type and have smaller leaves, but they are hardier and capable of self-seeding and maintaining themselves with little care in a way that the more highly-bred types are not.

It is possible that with selective breeding *dulce*-like and celeriac-like forms adapted to colder climates and lower-maintenance gardens could be bred from wild and leaf celery, and indeed I have already made a little progress on this in my garden, simply by starting with a wide range of types and selecting the resultant crosses with the best vigour and thickest stems. Wild celery has hollow stems but solid stems like those of *dulce* can be selected for. Some strains in my garden also show signs of perennialism, growing back again after seeding.

One group of *secalinum* celeries is the Chinese celeries, also known as kintsai (which is just an old spelling of the Chinese word for celery, qíncài or 芹菜) or Nan Ling celery. Chinese celeries have generally undergone more selection than their Western counterparts. They tend to be more delicate and more colourful, and not quite so hardy. Western strains are sold as herb, leaf or cutting celery. There's also a Dutch heirloom cultivar usually marketed as 'Par-cel' or occasionally 'Zwolche

Leaves of different strains of celery

Leaf sections of different strains of celery

Krul'. It's a dead ringer for curly-leaved parsley and there's a lot of confusion on the internet as to which it really is, but the smell quickly gives it away as a celery.

In the kitchen I use leaf celery as something between a vegetable and a herb. The strong, aromatic flavour is an excellent addition to winter soups, stews and leaf sauces. Together with onions and carrots it makes the French *mirepoix*, often used as a base for sauces and soups. The ingredients are fried slowly, which sweetens them rather than browning them. In summer the young flower shoots can be used in stir-fries and the seeds make a wonderful spice. As with most Apiaceae, the

tender new shoots are best, but with celery older shoots are pretty good up to a point too. Herb celery puts on most of its leaf growth in autumn, winter and early spring, meaning that leaves are available when not so many others are.

Although the ancestral celery is a marshland plant, I find it unfussy, growing in any part of the garden it can spread its seeds to. It will also take a degree of shade.

As well as *Apium graveolens*, there are a number of perennial celeries from the southern hemisphere, including *A. prostratum* and *filiforme* in Australia and New Zealand and *A. australe* in South America. These three are all closely related and all have been

Herb celery

recorded as being used in local foraging traditions. Like northern wild celery they are salt marsh plants, but in Patagonia *australe* is sometimes called *apio silvestre* (wood celery) and is recorded as growing in "subantarctic forest and … Patagonian steppe near … streams and rivers",[5] suggesting that it would do well in a forest garden environment.

Hogweeds and cow parsnips
Heracleum species

APIACEAE

Talk about giving a plant a bad name! The hogweed (*H. sphondylium*) in my garden is neither a weed nor for the hogs; it is a valued vegetable. It does need to be handled with some care though.

Hogweed has some pretty knowledgeable fans. Roger Phillips, author of *Wild Food*, describes it as "Unequivocally one of the best vegetables I have eaten",[6] while Margaret Lear of Plants with Purpose, writing in *A Handbook of Scotland's Wild Harvests*, calls it "an epicurean vegetable for the hungry gap".[7]

Like many other members of the carrot family, the best part of hogweed is the young leaf shoot, picked before the leaves have properly unfurled. The tastiest way of eating them is to sauté them in butter, adding small amounts of water as needed to prevent burning, until they develop a melting texture and a slightly caramelised taste. A variation is to add a little stock after frying, then cover and simmer for another 10 minutes to braise them. One of my favourite uses of hogweed is to make soup. I slice up the shoots to 1cm (0.4in) lengths and cook for 10 minutes in some stock along with some wild garlic leaves, then use a hand blender to blend it to a smooth consistency. The result is very creamy, with a delicate but distinctive flavour. Apart from this, it is a useful vegetable in any recipe where it will be cooked through, such as stews, curries and pasta sauce. I have also found it very tasty cooked with other shoots in tempura.

Hogweed leaf stems are covered in tiny hairs which I usually prefer to rub off before cooking. As they get older they get stringier and it can be worth peeling larger shoots. They store well in the fridge for a few days; in my experience they store much better if dry, so best to leave cleaning them until you are ready to use them (but see below). If blanched in hot water for a couple of minutes they will freeze well for later use.

Other parts of hogweed that can be used include the immature flower heads, which come neatly wrapped in papery bracts, and the seeds. The flower heads are best while still inside the bract although they remain edible until the flowers

Hogweed leaf stems

Hogweed flowers at various stages of opening

themselves open. The dried seeds have a wonderful citrusy aroma which makes an excellent addition to a spice mix. Even more exotic uses of hogweed have been reported; the Plants For A Future website says that the leaf stems are tied in bundles and dried in the sun until a sweet substance resembling sugar forms on them. Professional forager Miles Irving reports that the original borscht, now a pickled beetroot soup, was made from lacto-fermented hogweed leaves.[8] The roots have been reported as edible, but the ones I have tried have had a very unpleasant flavour.

Unfortunately, hogweed does have one major downside. The sap contains toxins called furano-coumarins which will cause nasty burns if they get on your skin in bright sunlight. I know this from painful personal experience after getting some of its sap on my arm while strimming in hot, sunny weather (we aren't cursed with such weather too often in Aberdeen, so I wasn't aware of the risk at the time). The affected area formed a wound that took months to get better since every time it healed and the scab came off the new skin underneath was damaged again. Native hogweed isn't as bad as the invasive giant hogweed (*H. mantegaz-zianum*), but it still warrants caution.

As bad as this sounds, a little care is enough to avoid getting the sap on your skin in strong sunlight and I have never had any problems while handling it for food preparation. I always make sure to wash my hands immediately after cutting it up but that is all. Furanocoumarins are also found in more usual vegetables such as celery and parsnips and citrus fruits like grapefruit. In grapefruit they are famous for inhibiting the action of certain drugs, so it is quite possible that this might also be true with hogweed. Furanocoumarin levels are generally increased by damage and during storage, so it may be sensible to use your hogweed fresh and trim off the ends of the shoots. Quite how much furanocoumarin is in British hogweed is uncertain. Plants For A Future suggest that the subspecies *sphondylium* and *sibirica* (the only two listed on the Euro+Med plant database as being present in Britain) are not phototoxic but I would be sceptical about that and would treat all hogweeds as guilty until proven innocent.

All in all, hogweed may require some care but I would rather get to know the dangers of plants and then use them safely than live in ignorance and fear of them. The reward is the distinctive taste of a vegetable that well deserves the praise heaped upon it by wild food enthusiasts.

There seems to be some doubt as to whether hogweed is biennial or perennial. All I can say is that the ones in my garden are perennial and long-lived. Given this, it is best to prevent them from seeding as they will do so in abundance.

In North America there is a close relative of hog-weed (*H. lanatum*), variously called hog parsnip, Indian celery or Indian rhubarb. Nancy Turner reports that it was used by virtually all First Peoples in British Columbia.[9] As you can guess from these names the stems are used again. Sam Thayer detects a distinct difference between the leaf stalk and the petiole, both of which should be peeled before eating: "The flavor of the petiole is sweet and fruity, reminding me of celery, dill and lemon. Some people like them dipped in sugar or chopped in little cross sections and added to fruit or green salad. The flavor of the flower stalk is less sweet

Native hogweed in my garden

– decidedly different but still good."[10] This distinction is also recognised in British Columbia, where the Stl'atl'imx (First Nation) called the leaf stems 'woman's foot' and the flower stems 'man's foot'.[11] Cow parsnips can also be used in all the same ways as hogweed but, despite the name, the root is still not much good.

In northern China there is another hogweed, *H. moellendorffii*, which Hu Shiuying reports is used similarly,[12] and even South America has one, *H. tuberosum*. According to Plants For A Future this last 'hog parsnip' finally has pleasant tasting roots.

Licorice-roots and small lovages
Ligusticum species

APIACEAE

If you don't have room for the monster herb that is lovage (p.144), perhaps you could try Scots lovage (*L. scoticum*). It is a low-growing herb with a similar, but milder and sweeter, flavour that grows on beaches and rocky shores around the coast of Scotland (and much of the rest of the North Atlantic

too). One of the nicest meals I ever had was of fresh mussels and Scots lovage, all picked off and cooked on the rocks on a canoe trip around the far northwest of Scotland. Despite this maritime distribution in the wild, it grows very happily as a woodland edge species in the forest garden. You can use leaves of any age as a herb or the young leaf shoots as a vegetable. The flavour goes especially well with tomato in a pasta sauce. The seeds can be used as a spice, with a flavour more akin to fenugreek or cumin.

Scots lovage has lots of relatives, many of which find a traditional use in the part of the world where they grow. The roots of Sichuan lovage or chuānxiōng (*L. striatum*) are esteemed as one of the 'eight precious items' in Chinese *bupin*. It is one of the first recorded aromatic herbs in Chinese literature.[13] Canby's lovage (*L. canbyi*) has a similar place in British Columbian culture, where the root, which Nancy Turner describes as having an "unforgettable sweetish aroma" is used as a "smoking condiment" by several First Peoples.[14] According to Dave Jacke and Eric Toensmeier, authors of *Edible*

Leaves and roots of Scots lovage

Not, as it turned out, chamnamul

Forest Gardens, the leaves, shoots and stems all have a more conventional culinary use too.[15] Many North American ligusticums are called licorice-roots. Hulten's licorice-root (*L. hulteni*) is a North Pacific version of Scots lovage and may in fact be the same species. Unfortunately when I have tried the root of Scots lovage it has had an acrid taste and a woody core like skirret.

Chamnamul
Pimpinella brachycarpa

APIACEAE

Chamnamul is highly esteemed in Korea. Its name means 'true' *namul*, an adjective often applied to the best in a group. Given how many *namul* there are in Korean cuisine, this is praise indeed! In South Korea, the young stems and leaves are eaten fresh or blanched as a *namul*. In North Korea it is popular as a kimchi ingredient. According to the Korean food blog Bburi Kitchen,[16] real chamnamul (true true namul?) is hard to find even in Korea, as the similar-looking *Cryptotaenia japonica* or *mitsuba* (see Chapter 9) is often offered under that name. Indeed, this is what happened to me: seeds that I ordered from Korea came up identical to my mitsuba and clearly different to chamnamul, which has a purplish tinge to its stems and more lobed leaflets. Since the native range of chamnamul extends to the Russian Far East it is likely to be hardy. It is described in the *Flora of China* as a plant of forest margins and river banks,[17] but Bburi Kitchen says that it is harvested from the deep forest, so it would be worth trying in both situations. Just make sure that what comes up is the real thing![18]

Parsley
Petroselinum crispum

APIACEAE

Parsley is criminally underused in British cuisine, relegated to use as a seasoning or, worse still, as a disposable garnish to pretty up a plate but rarely to eat. It deserves much more than this, being a tasty, hardy, biennial umbellifer that will maintain itself in your garden by seeding with little trouble. Like many of its relatives, it provides young shoots that go wonderfully in a stir-fry – even my parsley-hating partner likes them this way. Like far fewer of them, it produces them all winter long. You can combine parsley and celery shoots with garlic and broad beans to make 100% home-grown falafel in the depths of a Scottish winter. There is also a fleshy-rooted cultivar called Hamburg parsley that can be used as a self-seeding root vegetable. I have noticed that at least some parsley plants aren't

Curly-leaved parsley

entirely aware that they are biennials and grow on for more than their allotted two years, especially if someone eats all their flower shoots.

Self-seeding roots
Various species

APIACEAE

I once took over an allotment that had been neglected for many years. The previous plotter clearly had a thing for giving household waste a second life in the garden, and I dug out what felt like tonnes of old carpet, milk bottles and CDs,

but I was surprised to also unearth a generous harvest of big, fat, healthy parsnip (*Pastinaca sativa*) roots, many weighing in at over a kilo. This was all the more impressive since much of the plot was covered in rosebay willowherb and nettles: indicators of good fertility but no mean competitors for a feral root vegetable. According to Sam Thayer, parsnips have also widely naturalised in North America,[19] the only conventional root vegetable that I can think of to have done so.

Since I had always struggled to grow parsnips the proper way I took the hint and have been allowing mine to self-seed ever since. All the work I do for my parsnips now is to thin the seedlings out a little in spring and dig up the big roots in autumn and winter. The best roots of all are not eaten but replanted to run to seed and produce the next generation. It sounds ideal but there is one caution I must pass on: parsnip sap has the same effect on the skin in bright sunlight that hogweed does, so I make sure that my seeding plants are well back from paths.

It's an approach that can be taken with other root crops in the celery family. The seeds of many Apiaceae store badly in dry conditions and need winter cold before they will germinate. The easiest way to give them the conditions they need is to let it happen naturally. All it needs is for some plants

Self-seeded parsnip and 40cm ruler

Parsnip seedling

Turnip-rooted chervil seeds

to be left to seed and a degree of soil disturbance. It also helps to learn to recognise the plants as seedlings so you don't take them out as weeds!

Other roots that I grow this way include Hamburg parsley and a little-known vegetable called turnip-rooted chervil (*Chaerophyllum bulbosum*, or TRC for short). TRC roots have a delicious, chestnut-like flavour, but are little known because of the difficulty in growing them. They seem particularly intolerant of dry storage, so you need to

get fresh seed early in autumn and sow it, wait for the plants to sprout in spring and then leave the roots in the ground until the next winter as a period of cold improves their flavour. Allowing them to self-seed short-cuts all this difficulty, but brings its own problem: by this time they have died down so thoroughly that they are impossible to find. I harvest most of my TRC in the short window when they are sending up new shoots in spring, giving their location away.

Biennial spices
Carum carvi and others

APIACEAE

The Apiaceae need a whole spice rack to themselves. Besides those that need a hotter climate – like aniseed and cumin – and those mentioned elsewhere in the book – the perennials spignel, fennel, hogweed and Scots lovage and the biennials celery and alexanders – there is a further group of biennials that can be encouraged to self-seed in the same way as the root crops in the last section.

Turnip-rooted chervil

197

Caraway plants emerging in spring

These include dill (*Anethum graveolens*) and coriander (*Coriandrum sativum*) in the open sun and chervil (*Anthriscus cerefolium*) in shade. Caraway (*Carum carvi*) grows well on the forest edge, often taking three years rather than two to come to seed. The seeds of all can be used as spices and the leaves as herbs. Caraway is rare in the wild in Scotland but in *Around the World in 80 Plants* Stephen Barstow reports that it is foraged widely as a spring green in Scandinavia and that karvekålsuppe (caraway soup) is a popular rite of spring.[20] The roots are also very pleasant and Stephen is experimenting with breeding for larger ones. If you want to grow caraway for its leaf shoots, it is worth finding the cultivar 'Polaris', which holds its leaves upright and free from the soil. Most forms press their first leaves close to the ground, which helps them to compete with weeds but leads to dirty shoots.

Caraway in flower to the right of the path. Its fellow umbellifer, sweet cicely, presents a contrast on the left.

Sweet cicely (top), caraway and angelica seeds

Wild sarsaparilla
Aralia nudicaulis

ARALIACEAE

Wild sarsaparilla is like a miniature version of udo or American spikenard. The young leaf shoots can be fried, steamed or used as a pot herb. The boiled roots are nutritious and pleasant tasting. According to Daniel Moerman's *Native American Food Plants*, the fruit, which my plants haven't produced yet, has been used to make jelly and wine.[21]

Wild sarsaparilla

Hostas
Hosta species

ASPARAGACEAE

I sometimes struggle to persuade people that such a well-known ornamental plant as hosta is edible, but it has a long history of use in the Far East. It's not the first crop to be treated this way: the tomato was regarded as poisonous for around two hundred years after its introduction to Europe and North America.[22]

The best part of the hosta is the rolled up cigar of leaves as they emerge in the spring. Stephen Barstow calls them 'hostons', like the chicons of chicory. They come in many sizes, according to species and variety. Small ones are delicious stir-fried or wilted and dressed as *namul*. Chunkier hostons are better boiled briefly and used as a vegetable.

Hostons are best cropped by gripping them firmly near the base and snapping ones off the edge of the clump. If you can snap them off right at the base they will hold together as a whole instead of falling apart into individual leaves. The short leaf scales around the base are more bitter than the larger leaves so they are worth removing. It is possible to harvest the whole first flush of leaves of an established plant without killing it: ornamental hosta growers will sometimes 'mow' their plants to get a second flush of fresh, attractive leaves.

Leaves and petioles can also be used as a pot herb, although not the most exciting one, after

Large and small hostons. The large ones will be boiled, the small ones stir-fried.

Hostas going to waste in a flower bed, Royal Botanic Garden Edinburgh

Hostas 'Sagae' (left) and 'Empress Wu' (right)

Hosta shoots emerging; more mature ones in the background

unfurling, so long as they remain tender. The flowers and flower buds are also edible: the Montreal Botanical Garden lists all species as edible and *H. fortunei* (a synonym of *H. sieboldiana*) as the tastiest.[23] I confess that I don't like them though.

Edible species include *Hosta clausa, longipes, longissima, minor, montana, nigrescens, plantaginea, rectifolia, sieboldiana, sieboldii, tardiva, undulata* and *ventricosa*.* There are many cultivars which may be complex crosses between the species. Hostas are unusually mutable and new mutations are always cropping up and being selected by breeders. I haven't heard of any adverse effects from eating any hosta species or cultivars but it is always sensible to be cautious when trying one for the first time. Ones in my garden include 'Sagae', a cultivar of *sieboldiana* grown widely for food in Japan, 'Empress Wu', one of the largest, and 'Alice in Wonderland'.

In their native habitat many hostas are forest plants, but I find that in my climate they don't flourish in deep shade: a spot in sun for part of the day and shade for part is just fine. Blue-green varieties are generally more shade-tolerant and variegated ones less so.

Triteleias
Triteleia species

ASPARAGACEAE

A final genus in the asparagus family that might be useful on the forest edge is *Triteleia*, also known as the brodiaes, tripletlilies or cluster lilies. I haven't grown them to maturity yet myself, but Nancy Turner records that the corms of *T. grandiflora* and hyacinthina, both native to British Columbia, were used by First Nations there.[24] *Triteleias* sold in the UK are usually the less hardy (but also edible) *T. laxa*, Ithuriel's spear.

Daylilies
Hemerocallis species

ASPHODELACEAE

Daylilies have been described as 'the perfect perennial', due to their brilliant colours and all round ease of growing. They tolerate both drought and frost and thrive in many different climate zones and soil conditions. They are vigorous perennials that last for many years in a garden and see off most weeds. They like full sun but will also grow

---

* The species I have eaten regularly myself are *H. minor, sieboldiana* and *longipes*. Crawford (2010) lists *H. crispula* (syn. *sieboldiana*), *longipes, montana* (syn. *sieboldiana* var. *montana*), *plantaginea, sieboldii, sieboldiana, undulata* and *ventricosa*. Plants For A Future add *H. clausa, clavata, longissima, nigrescens, rectifolia* and *tardiva* and list no known hazards for the genus as a whole.

Daylily development

with shade for part of the day. As if all that wasn't enough, they are really nice to eat too.

Many parts of the daylily plant are edible. The young leaves can be used as a pot herb or, in small amounts, as a salad. They emerge in spring in bundles like leeks and I sometimes thin overcrowded plants in spring and use these individual shoots.

However, the main attraction of the daylily, in culinary as well as ornamental terms, is the flower. According to Hu Shiuying, the best time to use it is just before the flower bud opens out into the flower.[25] On opening, and then even more on drying, the plant takes nutrients out of the flower, back into the plant. Buds can be cured by heating and drying or used straight away: frittered, stir-fried or chopped and used like an onion. After opening, the flower is less nutritious but still a great visual addition to dishes. The key to getting the best out of them flavour-wise seems to be frying, which caramelises the sugars in the petals. They have a useful ability to thicken soups and stews: a simple but attractive miso soup can be made by frying a handful of daylily flowers, then adding hot water, cooking for a short while and adding miso.

After a short while (hence the name daylily) the flower withers and dries. It is then still less nutritious, but retains its ability to thicken, and dried flowers are often collected and stored for this purpose. Some people also use the flowers in salads; personally I find they have an unpleasant aftertaste and slight burning sensation and only ever use them cooked. Flower colour may have an influence on taste. The daylily breeder Brian Reeder,* who breeds cultivars for edibility as well as appearance, writes on his blog that "some of the darker anthocyanic pigments found in *Hemerocallis* hybrids also gives a stronger, somewhat more astringent flavor to the petals, while pale yellows, lavender, pink and near white are often very light and sweet."[26]

Finally, the roots, which on some species form tuber-like thickenings, are also edible. The forager and chef Hank Shaw, who writes the 'Honest Food' blog, describes them as "quite possibly the best tubers I've ever eaten".[27] I rarely use them because they are not produced in great quantities and it involves digging up the plant, but if you are dividing them or are lucky enough to live in one of the places where daylily has become an invasive weed, take full advantage.

There are many species of daylily. Plants For A Future list over 20 and only one, *H. forrestii*, gets anything less than a four-star rating for edibility. As with the hostas, many daylilies that you might encounter do not fit strictly into any one species

* A clear case of what *New Scientist* magazine calls nominative determinism. Brian's blog is called Daylily BReeder.

Hemerocallis fulva 'Kwanso' with double flower

as they have been hybridised widely and many are listed only as *Hemerocallis* and their cultivar name. I would be a little more cautious trying new cultivars of daylily than hosta, as they are more chemically complex. According to Brian Reeder, who must have tasted a large number of novel cultivars, "I have found that some of them are good to eat, while others have odd, sedative effects, and others are almost laxative in nature."[28] Personally I have never noticed any ill effects from cooked daylilies and it should be remembered that they have been used in China for thousands of years, but care would be advised when trying any cultivar that is new to you.

Different daylilies flower at different times and having a range of species and cultivars ensures a succession of flowers at different times of year. I grow *H. citrina*, *dumortieri*, *esculenta*, *fulva*, *lilioasphodelus*, *middendorfii* var. *exaltata* and *minor* and a range of cultivars, including 'Whichford', 'Burning Daylight', 'Franz Hals', 'Yellow Moonlight', 'Pink Damask' and 'Cream Drop'. You can also find double cultivars of daylily which have the culinary advantage of being chunkier: *H. fulva* 'Kwanso' is one that I grow. Two common ones that I have found rather disappointing in terms of size and yield are 'Stella de Oro' and 'Corky'. Many new cultivars are becoming hard to recognise even as daylilies. The trend

in breeding seems to be for more open flowers, with petals curved back hard – pretty much the opposite of what you want for cooking. Species and older cultivars are generally better.

Growing daylilies is easy. They do best in a moist, fertile soil in sun or semi-shade. There is apparently a daylily gall midge (*Contarinia quinquenotata*) which can lead to distorted flowers. Fortunately I have never seen it in my garden. Slugs are fond of the young growth. This isn't a problem with established plants but new plants are worth protecting when first planted out.

Burdocks and thistles
Arctium and *Cirsium* species

ASTERACEAE

If you grow burdocks, you'll soon find out why they were the inspiration for the invention of velcro. Their hooked seeds hitch a ride on anything vaguely woolly, including the gardener's jumper or hat, to aid their dispersal. Often they fall off eventually, but their ideal is for their host to die with the passengers still in place, giving a planting spot with the rich, fertile conditions that they like best. So far I haven't given them the satisfaction.

There are two species that you can grow, both of which yield different products at different points

Young burdock plant

in their biennial lifecycle. The greater burdock, *A. lappa*, is probably best known by its Japanese name of gobo. Lesser burdock, *A. minus*, is a native Scottish wildflower. As its name suggests, it is smaller than gobo, but its flower stems will still top six feet in fertile conditions. I grow both in my garden and harbour hopes that they will hybridise. The first-year product is the root. This can reach up to a metre long, so digging one up is quite a project. I like them best cooked the Japanese way as kinpira: julienned and then braised with soy sauce, mirin and sesame oil. They can also be used as a filling for maki rolls or in stews. According to Moerman's *Native American Food Plants*, the Iroquois dried burdock roots for use in winter.[29]

In the second year, the part to use is the young flower stem. With its bitter skin peeled off and cooked, it has an artichoke taste. If you leave it alone at this stage it will produce the magnificent mature flower stem and those hitch-hiking burrs. The nectar from the flowers attracts a range of butterflies and the seeds, which are produced in large quantities, can be sprouted and eaten. When processing the seeds, look out for the large quantity of irritating hairs that come out of the seed head along with them.

The plume thistles of the genus *Cirsium* are closely related to burdocks and some can be used in similar ways. In season, I forage the flower stems

Burdock flowers

Dry burrs for seed collection

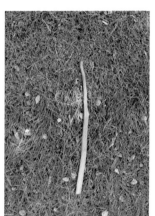
Peeling a burdock flower stem

Cabbage thistle

Cabbage thistle flower

of the native spear thistle (*C. vulgare*) for the sweet, juicy pith, but I wouldn't encourage them in my garden as they are ferociously spiny. Cabbage thistle (*C. oleraceum*) is a more agreeable customer, with soft, unspiny foliage that can be used as a pot herb and shoot vegetable. Unlike its cousins it is a perennial which spreads slowly to form a clump, growing to a little over a metre in height.

Asters
Various species in tribe *Astereae*

ASTERACEAE

The genus *Aster* was recently broken up by botanists into a slew of new genera, mostly with unlovely names like *Symphyotrichum* and *Doellingeria*, but most of them continue to be known as asters in their common names (and in Latin by those unwilling to give up a blessedly short, memorable name).

The group seems quite prone to confusion in the nursery trade: not so important if you are just after a pretty flower, but more significant when you are planning to eat them. For instance, I grow (and eat) a cultivar called *A. ageratoides* 'Ezo Murasaki'. I originally bought it as *Aster scaber*, but when I had a visit from Stephen Barstow, who is a big fan of the real *A. scaber*, I discovered that it was no such thing. When I went back to the nursery I bought it from they told me that they had discovered that it was actually *Aster trifoliatus* subspecies *ageratoides*. Now *Aster trifoliatus* doesn't even exist, but *A. trinervius* subspecies *ageratoides* is a synonym of *A. ageratoides*, which my plant fits the botanical description of. Buyer beware! Fortunately, given all this, while many asters aren't particularly worth eating, I haven't come across any risks in the former genus as a whole, so nibbling on the wrong aster is unlikely to kill you.

The asters are mostly plants of scrub, wood margins and open woods, often preferring a moist soil. The ones of interest are all rather sprawling, medium-sized herbaceous perennials. They grow in expanding clumps: some; like *ageratoides*, expand fast enough that some care is needed in placing them but most are quite well behaved. They mostly flower very late in the season, giving rise to some poetic names. The New England aster is known as 'the last rose of summer' and asters in general are often called Michaelmas daisies, after the festival celebrated on the 29th of September, now named for the Archangel Michael but originally honouring

Leaves of *Aster ageratoides* 'Ezo Murasaki'

A hoverfly enjoying a chamchwi flower

the Celtic sun god Lugh. When, as I do, you harvest the flower shoots heavily, they may not even flower until the start of winter, justifying a final name of 'frostflower'.

The nation most keen on eating asters is Korea, where the word *chwi* refers to all sorts of plants in the aster family. Used as vegetables, they are *chwinamul*. The most esteemed of these is *Doellingeria scabra*, the former *Aster scaber*, which is known as *chamchwi* or 'true aster'. Originally collected as a mountain herb, it is now widely cultivated and seeds can be bought from Korean seed suppliers. The leaves are used in salads, stir-fried or blanched as *namul*. They are used

after stir-frying as a wrapping leaf (*ssam*) to wrap strips of meat and other vegetables, brined to make kimchi or pounded with rice to make rice cakes.[30] Other hardy asters used in Korea include *Aster tataricus* (Tatarian aster or *gaemichwi*),[31] *Aster glehnii* (Ulleung Island aster) and, for a real tongue-twister, *Miyamayomena koraiensis*.[32]

Aster ageratoides is also from Asia, with a natural distribution including China, Korea, Japan and the Russian Far East. I don't know of any traditional use of it but I find that the flower shoots, which are produced in great abundance, stir-fry nicely. It is available in a range of ornamental forms. If you like more of a challenge, you might want to try tracking down *Galatella hauptii*, a steppe aster from Central Asia which receives a mention (as *Aster*) on Plants For A Future.

From Asia we move to North America, where most of the asters are now in the genera *Eurybia* and *Symphyotrichum*. The New England aster (*S. novae-angliae*) is a tall herbaceous perennial that has become a popular garden plant, bred into many different varieties. One website describes it as having "very mild, mint-like scent".[33] Timothy Coffey mentions the use of the blue wood-aster (*S. cordifolius*) as 'greens',[34] and the bigleaf aster (*E. macrophylla*) was used by the Algonquin and Ojibwa[35] and is popular on US and Canadian foraging sites for its young leaves. I have yet to try any of these but I suspect that the flower stems would be useable in the same way as the Asian species.

All the asters dry quite easily and can be reconstituted for later use or crumbled into winter soups and stews for flavour.

Gomchwi
Ligularia fischeri

ASTERACEAES

This one's for the bears! It's another Korean *chwinamul*, and its name literally means bear aster. This is always a good sign as our hairy friends seem to have quite similar tastes to humans, and indeed

Gomchwi flower

Gomchwi (*Ligularia fischeri*)

gomchwi is one of the most popular *sannamul* in Korea – to the extent that it has become a cultivated crop as well as a wild mountain vegetable. It is cooked in all the most distinctive Korean styles: steamed and dressed as a *namul*, fermented in kimchi or used as a *ssam* to wrap other foods.

Plants from different sources can vary considerably. A plant I grew from seed received from Stephen Barstow (who obtained his original seeds from Korea) has light green leaves and a mild flavour. Two other plants received from other sources and originating elsewhere in the species' range have darker, thicker leaves and an unpleasantly strong taste. If you want to grow it, I'd suggest getting seeds from a Korean seed company.

Other species of *Ligularia* have become popular garden plants in Scotland, especially the cultivar 'The Rocket'. Gomchwi shares with these a tolerance for some shade and an alarming habit of completely collapsing if it becomes too dry – although it will make a full recovery if watered in time. A moist, shady, forest edge spot is perfect for it.

Comparison of mild and strong *Ligularia fischeri*

Shidoke
Parasenecio syn. *Cacalia delphiniifolia*

ASTERACEAE

Shidoke is a woodland sansai which the blog 'Shikigami' describes as having a "fantastic and unique flavour" when boiled very briefly.[36] I haven't managed to obtain it yet but it sounds very promising.

Butterburs
Petasites species

ASTERACEAE

Butterburs are strange, primeval-looking plants, although deep down they are just another daisy. They flower before they come into leaf, and since they can spread over large areas by underground runners this can be an impressive sight. The leaves then look rather like rhubarb, but without a central point that they all grow from. Many species are edible, but the best (because it is the largest and least fiddly) is Japanese butterbur, *Petasites japonicus*, or fuki.

Traditional cooking methods are designed to remove the toxic, bitter alkaloids found in the raw plant. There are two edible parts: the flowers, called *fuki no to*, are picked when still in the bud, and the leaf petioles are picked while young. Petioles are boiled for half an hour in water (although some recipes use much less than this), then they are cooled and the skin is peeled off (it is thin and comes away easily by hand at this point). Like gobo, they are good at taking up the flavour of seasonings and they can be cooked in the same way. The flowers are boiled and then squeezed to remove the water before being used in tempura or cooked with egg, broth or miso.

Like other butterburs, Japanese butterbur is quite spready, but its large leaves make it easy to pull out of areas where you don't want it. Don't plant it on the border of your plot though, or you

Fuki no to, Japanese butterbur flowers

Japanese butterbur

Butterbur stems fried in sesame oil with soy sauce and mirin

207

will get a reputation as a bad neighbour. It likes a moist, rich soil and will stand a fair bit of shade. The leaves grow to about 0.6m (2ft).

Sochan
Rudbeckia laciniata

ASTERACEAE

Some plants a gardener seems fated never to get. I learned about sochan a few years ago from Sam Thayer's book *Incredible Wild Edibles*[37] and decided that it sounded ideal for a forest garden. It is a common wildflower through much of North America, typical of streamside forests and meadows, floodplains, carr (waterlogged woodland), fence rows and forest edges. Flower growers will probably recognise the Latin name: it is one of the coneflowers, like black-eyed Susan. Since then I have been trying to get a plant, but seeds from various sources have failed to sprout and plant orders have fallen through for one reason or another. The latest was cancelled because of the lockdown for COVID-19.

This is a pity, because sochan sounds delicious. It is one of the five most important wild foods of the Cherokees, and Stephen Barstow rates it amongst his 10 favourite perennial vegetables, writing on *Edimentals,* "wow, I hadn't expected it to taste that good, slightly sweet and aromatic".[38] Any tender aerial parts can be used. You can steam or boil them as a vegetable, or cook them Southern-style, fried in pork fat and drizzled with a vinegar-based dressing. Thayer peels larger shoots and uses them raw. I am determined to defy fate and try sochan for myself sooner or later!

Scorzonera
Scorzonera hispanica

ASTERACEAE

One last member of the aster family for now: scorzonera is a perennial relative of the salsifies and goatsbeards, although unfortunately its flower heads aren't nearly as nice to eat. It is sometimes

Scorzonera flowers

grown for its first-year roots (older ones are edible but woody and tasteless), but for me the outstanding part is its leaves, which are produced early and abundantly in spring. These are mild-flavoured, working equally well as a pot herb or salad. Their main drawback is a faint white, woolly covering on the leaves, which spoils the mouth-feel in a salad. Luckily, this varies considerably from plant to plant and some are almost completely free of it. With a little selective breeding I'm sure it would be possible to produce a line of 'salad scorzonera'.

Scorzonera leaves

Land cress leaves and flower shoots

Land cress running to flower

Wintercresses
Barbarea species

BRASSICACEAE

The wintercresses, also known as yellowrockets, are a group of low-growing plants with a strong cress flavour. The biennial species *B. verna* is the one you are most likely to find in seed catalogues, where it is usually called land cress or American land cress, despite the fact that it originally comes from Southern Europe. The 'land' part is an attempt to sell it as a dry-ground substitute for watercress. It is true that land cress is far easier to grow than watercress, being much hardier and not needing wet soil, but it is only a partial substitute. I find the taste of the leaves much harsher than those of watercress. I use them only very sparingly in salads, but they do make a very nice alternative to watercress soup, blended into a potato soup at the last minute. The part I use far more is the flowering stems which grow in spring. They are milder flavoured and very large in proportion to the plant. I mostly steam or stir-fry them, like peppery broccoli sprouts.

The native Scottish species is common bittercress (*B. vulgaris*). It has been grown in Scottish gardens since at least 1862 when it was mentioned by CP Johnson in *Useful Plants of Great Britain*.

Although it is also described as a biennial I often find that plants come back for at least another year after flowering. There is a variegated form which, unusually, grows true from seed; the variegation carries into the flower stem, making for a very exotic-looking vegetable indeed. Wintercresses are very hardy; both common bittercress and land cress grow slowly throughout the winter in my garden, although Stephen Barstow says that in Norway his don't.[39] The Latin name for the genus comes from the availability of the leaves on St. Barbara's Day, the 4th of December. Both self-seed effusively and need some weeding out where you don't want

Variegated bittercress in winter

them. Fortunately they don't seem to be affected by any of the diseases of the brassica family.

There is a third species, a genuine American wintercress, called *B. orthoceras*. It is the hardiest of the three and largely replaces the imported *verna* and *vulgaris* in the boreal forest regions and Rocky Mountains in North America.[40]

Kales
Brassica oleracea

BRASSICACEAE

Brassica oleracea is the plant breeder's most impressive advert. A single species has been selected for different parts to produce vegetables as varied as cauliflower, cabbage, Brussels sprouts, kohl rabi, broccoli and gai lan. This presents challenges for seed savers. Unless you want to try producing a novelty – Brussels cauliflowers say – the different vegetables need to be strictly segregated to avoid cross pollination. This is less of an issue when you only want to harvest the leaves (known as kale or collard greens) and spring flower shoots (known as sprouting broccoli but actually produced by all types of brassica if you allow them). For these you can keep a self-seeding, self-maintaining population in true forest-garden style. Better still, you can breed in a degree of perenniality, which helps to keep superior or interesting cultivars distinct despite the shared gene pool.

Kales were so esteemed in Scottish gardening that a vegetable garden, in Scots, is called a 'kail yaird'. It's not hard to see why: kale is hardy, productive, tasty and nutritious and easy to collect seed from. With their winter leaves and spring shoots they played a vital role in filling the 'hungry gap' when little other palatable greenery was available.

Most modern kales are biennial and have to be regrown from seed every year, but there are a few perennial ones, such as Daubenton's kale, which exists in both regular and variegated lines. It is available from nurseries in Europe grown from

cuttings. In the US it is much harder to get hold of, but the cultivar 'Kosmic Kale' seems to be very similar.

Daubenton's gains its extra years of life by rarely, if ever, flowering. Individual plants are quite short-lived. They become more frost-sensitive with time and usually succumb in a hard winter around year five or six, but they have a trick up their sleeve. Stems arch over and root where they touch the ground. This process seems to reset the age of the plant and the kale survives as an ever-expanding circle of rejuvenated plants.

Fortunately for us, Daubenton's does break its vow of celibacy very occasionally. One such

Kale breeding in progress

Daubenton's kale stem with rooting points

flowering happened in the allotment of the Welsh gardener Graham D. Jenkins-Belohorska. Also fortunately for us, Graham shared the results of his good fortune and there is now an ever-expanding 'grex' of lines descended from his original crosses. In a Facebook post in 2018, Graham wrote:

"The descendants of a handful of plants growing in an allotment in a small Welsh village are spreading round the world and mixing with local landraces; and perennial kales are becoming far more diverse, and far less rare.

"So much potential was locked up in those first seeds, and it gives me genuine joy to see how that potential is being unlocked in the hands of others around the world.

"I had a little good fortune back in 2010 when Daubenton decided to flower for me. And had I been possessive of that fortune, all that potential could have been lost. But the simple act of sharing some seed has set all that potential free, and regardless of what happens to me and my own kales, I know I got to play a small part in the story of perennial kales and their current resurgence."

I have found in my own experiments that

Kale seed head

descendants of Daubenton can occupy a sweet spot between 'flower yourself to death' bienniality and 'No Sex Please, We're Perennial'. These plants flower without dying, meaning that you get the advantages of a perennial but can make crosses from it to breed perennial traits into your favourite kales. Indeed my kale population now includes perennial plants which clearly have cabbage and Brussels sprout from neighbouring gardens in their

Some different kale leaves

lineage, bringing extra sweetness and tenderness to the leaves. I have largely given up on deliberate breeding, merely selecting the most interesting plants from the ones that spring up around where I have planted them to grow and take part in the next generation of crosses. To a traditional grower this probably sounds like a recipe for rampant club root, but I have never noticed a Daubenton's kale bothered by it.

Besides the perennial kales, there is a kaleido-scope of biennial cultivars, with wildly different colours, leaf shapes, sizes and textures. 'Pentland Brig' is my favourite for flavour, and I also grow 'Ragged Jack', 'Nero di Toscana' and 'Sutherland Kale'. If you want to maintain a cultivar rather than stirring the gene pool, the best way to do it with kales in a small garden is to ensure that the plants of that breed flower at a different time from the others. This can be achieved by stopping eating the flower shoots earlier or later than the rest!

The traditional Scottish use of kale leaves is in soup, and it's true that kale soup is delicious, especially blended with potato and some strong herbs or spices, but I also find them useful as a general-purpose pot herb, or you can salt and bake them for kale crisps. The flower shoots are even better. Simply steamed and served with a little olive oil, lemon juice and salt they are wonderful in both flavour and texture. They are a the-more-you-pick-the-more-you-get crop, from which you can take several harvests before finally leaving them to flower and seed.

Kales grow fastest in full sun, but they tolerate a fair degree of shade and grow more tender leaves as a result.

Mustard
Brassica juncea

BRASSICACEAE

Brassica juncea is almost as mutable as *oleracea*. The *integrifolia* group of leaf mustards includes the cut-leaved mizuna, crispy Texas mustard, large-petioled horned mustard and heading *dai gai choy*. Other groups have been selected for roots, stems and oily seeds. Its seeds make 'brown mustard', although the more familiar yellow mustard comes from the related *Sinapis alba*.

In my garden I use mustard as a fast-growing, self-seeding annual to make use of temporary space. The main ancestors of my mustard population are a cut-leaved variety that goes into salads and a large-leaved red Japanese mustard for steamed or stir-fried mustard greens. Fast-bolting or small-leaved plants are pulled up and go on the compost as a green manure. The best plants are allowed to

'Sprouting broccoli' from kale   A variety of mustard leaves

go on and seed for use as a spice and to provide the next year's generation.

Turkish rocket
Bunias orientalis

BRASSICACEAE

Turkish rocket is a good example of the fact that you should never be too fast to give up on a plant whose virtues aren't immediately obvious. When I first grew it I followed the advice on Plants For A Future: "the cooked leaves make an excellent vegetable". I'm afraid I can't agree. To me, the leaves have an odd bitterness which is capable of spoiling an entire dish. They are somewhat milder in the early spring but still not very pleasant. Eventually I discovered that the best bit is the immature flowering stems (rather like sprouting broccoli – I call them 'rockoli'). They have an unusual, slightly shellfish-like flavour that at first I found frankly disturbing in a plant, but I have come to like it and now look forward to the rockoli season keenly. They are one of my favourite *namul*, and nice cooked in a white sauce with a little cheese and mustard.

Even with this discovery it seemed a rather unproductive plant, producing only small and fiddly secondary rockoli after the first flush. More recently I have discovered that the remedy for this is to pick a good long stem along with the flower head. Not only does this give you a larger crop, it prompts the plant to produce another – and another – crop of large flower stems.

When Turkish rocket does flower it produces great clouds of little yellow blooms that are very popular with hoverflies. Unless you want to collect seed it is advisable to prevent it from seeding as it can self seed rather too effectively and become invasive. It is an extremely deep-rooted plant. I once dug a deep path next to mine and was still finding roots 1.2m (3.9ft) down. The grated roots have a pleasantly spicy taste like horseradish.

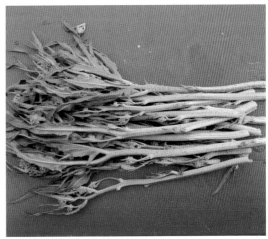

Rockolis

Turkish rocket flowers, perfect crucifers

Rocket
Eruca vesicaria

BRASSICACEAE

Rocket, or arugula if you prefer, presumably needs no introduction. It grows well as a self-seeder in the forest garden, with the added benefits that it will often overwinter with shelter from the east and that allowing it to self-seed helps to breed a frost-hardier population for your garden. You might be interested to know that Roman writers considered it an aphrodisiac. The cultivated kind is the sub-species or variety *sativa*. Perennial wall rocket (p.238) is a different species.

Radish
Raphanus raphanistrum

BRASSICACEAE

There is so much more to radish than the little round roots that they are traditionally grown for in Britain. You can eat the much larger roots of Oriental radishes like mooli, and the leaves, the 'broccolis' and the seed pods, all of which have a radishy bite. In China they are regarded as health-giving, reflected in the saying "Eating pungent radish and drinking hot tea, let the starved doctors beg on their knees"![41] I use them particularly in stir-fries and making kimchi. In Korea, the tops of radishes harvested after the first frost are cut off and hung up to dry, making a product called *shiraegi*. Shiraegi can be stored for up to three years, and are reconstituted by soaking for 24 hours and boiling for an hour.[42]

Plants grown for pods tend to sprawl, so I grow a patch and surround it with stakes and string like

Self-sown radishes

broad beans. They vary a lot, so there is potential to breed for longer, larger pods and to eliminate the rough skin found in some plants. Plants that come up in spring bolt quite quickly so you can only get tender roots by culling the youngest plants. Plants that sprout in autumn last longer and

Sprawling pod radishes

Fresh and dry radish pods

Sea radish flowers, from seeds collected on Easdale Island

make far larger roots, which will appear randomly in the garden if you allow them to self-seed. Radishes are unusual in that their pods do not split open as they dry, so they are easy to collect but awkward to extract seed from.

The cultivated radish is the subspecies *sativum*. There is also a wild radish called the sea radish (subsp. *landra*), which is much hardier and more robust than the domesticated variety, and which might have useful traits for a home garden population of self-seeding plants.

Horseradish flavours
Various species

BRASSICACEAE

The horseradish flavour can be found in many places around the cabbage family. The most recently fashionable is wasabi (*Eutrema japonica*), which is a fairly demanding plant to grow as it needs dappled shade and cold, running water over its roots. I manage to maintain a small plant without these conditions, but it isn't very productive. Horseradish (*Armoracia rusticana*) is exactly the opposite: if anything it can be too productive. I get round this by growing it in a relatively shady spot (it grows fastest in a damp soil in full sun) and using it as a pot herb and green manure as well as a spice.

Sea radish pods

If you buy 'wasabi' in Britain the odds are that it is in fact horseradish with green food dye in any case.

Ungrated horseradish roots have very little flavour, but once they are cut the damaged plant cells rapidly produce fiery mustard oil, a protective ploy against being eaten by less perverse species than ourselves. This volatile oil is easily lost, being broken down by exposure to air or by heating, so horseradish is generally added to dishes at the end of cooking and needs to be used quickly. Grated horseradish's shelf life can be extended by combining it with vinegar: horseradish sauce is made with both vinegar and cream. Even then, however, it only lasts a few months before losing its savour,

Horseradish leaves are very unusual. The first leaves each year come up deeply lobed, but later ones transition to a simple leaf.

Lady's smock

making it worthwhile to have your own plant. I also find that freshly grated horseradish has a sweetness and freshness that a sauce just can't match.

The heat of horseradish can be used in two ways: by matching it with other strong flavours or by using it to spice up blander foods, especially starchy roots and oily, creamy dishes. Classically it is used with strong-tasting meat or fish but there is much, much more to horseradish than this. Scrambled eggs with horseradish root grated into them are superb, or you can try adding it to a creamy pasta sauce at the end of cooking. Mashed potato is another dish that is enhanced considerably by horseradish flavour – if you substitute wasabi leaves for cabbage you can make an amazing variation on colcannon or bubble and squeak. Adding horseradish to soups – again at the end of cooking – gives quite fine control so you can add a very delicate horseradish flavour if you like or go for the burn. For real addicts like me, you can even make a horseradish salad dressing with fresh grated horseradish combined with olive oil, lemon and salt.

Before horseradish was introduced to the British Isles, two other, native plants were used for a similar flavour: Jack-by-the-hedge (Chapter 9) and dittander (*Lepidium latifolium*). Both have a superior flavour to my mind, but the roots are much smaller and more fiddly to prepare and I wouldn't

recommend growing them for this purpose. Both have edible leaves though, and those of dittander are especially interesting. Alone amongst the other plants, they have a strong horseradish flavour in their leaves as well as their roots, giving them the nickname 'wasabi leaf' in my garden. This gives an extremely easy way of adding wasabi flavour to everything from salads to cooked dishes (finely chopped – they are strong!). Pounding together grated horseradish, chopped dittander leaves and a little mustard gives a wasabi substitute for sushi that even looks the part. If you have too many dittander leaves they can also be used as a pot herb as they lose all their kick when cooked for a few minutes. Dittander can tolerate deep shade and is an invasive spreader if allowed to be, so grow it in a pot like mint.

Finally, the most decorative form of horseradish flavour is in the flower heads of lady's smock (*Cardamine pratensis*, also known as cuckoo flower), a native cress growing in damp and shady places. The leaves also have a horseradish flavour which is stronger than that in the flower heads.

Bellflowers
Campanula species

CAMPANULACEAE

There are many species of bellflower, all of which are technically edible, but having tried many I find myself using only a few. The giant bellflower is a forest shade plant, which we met in Chapter 9. Two that grow on the forest edge are Serbian bellflower (*C. poscharskyana*) and peach-leaved bellflower (*C. persicifolia*). They are very different plants. Serbian bellflower is low growing, with trailing stems that straggle through other vegetation or cascade over the edge of a bed. The stem tips, of which I harvest about 10cm (4in), have a pleasant nutty taste either raw or cooked. The flowers are edible but too small to pick for more than a garnish. Peach-leaved bellflower, on the other hand, is a very upright plant, with tough stems and leaves that are not worth trying to eat but large, showy flowers that are well worth picking for a salad. If you pick the whole flower you will find that the base is a little bitter but you will get a succession of new flowers. An alternative is to just tear off the corolla – the fused petals – which is better tasting but allows the plant to run to seed sooner.

Asian bellflower relatives
Adenophora, Codonopsis and *Platycodon* species

CAMPANULACEAE

A number of close relatives of the bellflowers are found only or largely in Asia, where they have both medicinal and edible uses. All the ones mentioned here are herbaceous perennials. The roots of many *Adenophora* species, known collectively as lady-bells, are reported to be edible, albeit often after boiling in two changes of water. The only one I have tried so far is *A. triphylla*. The roots are white and carrot-like in shape, and after a short while cooking they become soft and mild-flavoured. Unfortunately plants in this genus don't cope well

Peach-leaved bellflower

Fresh growth on Serbian bellflower (see also p.184)

with root disturbance, making a regular harvesting regime tricky.

There are many species of *Codonopsis* scattered around Asia, giving them the English name of Asia-bells. A number of them are grouped together as *dang shen*, a popular *bupin* in China, sometimes known as 'poor man's ginseng'.[43] The only one I can find recorded as being used as a regular food is *C. lanceolata* or *deodeok*, the roots of which are used both raw and cooked in Korean cuisine.

Platycodon grandiflorus, the only member of its genus, is known as balloon flower for its large flowers, the petals of which take a while to separate, resulting in inflated, balloon-like flower buds.

Codonopsis lanceolata flower

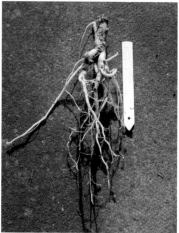

Codonopsis lanceolata root

It is also popular in Korea, where the roots, either fresh or dried, are one of the most common *namul* and *bibimbap* ingredients. Roots have to be sliced thinly and treated to remove a bitter flavour, by repeated soaking and draining and/or scrubbing with salt.[44]

Spiked rampions
Phyteuma and *Asyneuma* species

CAMPANULACEAE

Probably everyone knows the Brothers Grimm story of Rapunzel. A pregnant woman craves 'rapunzel', a variety of salad green, and her husband resorts to stealing some from the garden of a nearby sorceress. When he is caught, the sorceress says that he can take as much rapunzel as he likes, but must give her the baby in return when it is born. Much action, with the usual fairytale cast of princes and towers, ensues, and the sorceress comes to a bad end. She should have known better than to withhold the use of medicinal herbs from a forest garden!

The identity of the original rapunzel is unclear. The story overlaps with that of St. Barbara, whose saint's day is the 4th of December, a time when fresh greens would have been rare and particularly craved by pregnant women. The scientific name *Barbarea* for wintercress records this association

for one winter-hardy plant. It is possible that 'rapunzel' referred to a number of winter and early spring greens instead of a single plant. Nowadays, 'rapunzel' and the derived 'rampion' are reserved for the bellflower *Campanula rapunculus* and plants in the genus *Phyteuma*, also known as spiked rampions. I have to wonder if *C. rapunculus'* reputation for having edible roots derives from confusion with the spiked rampions. I have grown it many times with seeds from many sources (most of them explicitly sold as a root vegetable) and never found a hint of a fleshy tap root. Online pictures of rampion roots are invariably drawings or photos of what look like different species to me. Since some bellflowers (like the giant bellflower) do have edible roots it's also possible that fleshy-rooted rampions are just a vanished or rare selection and that most seed sellers aren't actually checking the roots of their stock plants.

With *Phyteuma* we are on much more solid ground. There are many species in Europe, of which black rapunzel (*P. nigra*) and spiked rapunzel (*P. spicatum*) are the hardiest. Both grow successfully (and hybridise) in Stephen Barstow's garden in Norway. Stephen has written a monograph on the rapunzels, available on his Edimentals blog.[45] The leaves are, as you might expect, edible as both a salad and a pot herb. The roots are described by

Black rapunzel (*Phyteuma nigra*)

A clump of *Tradescantia ohiensis*

Spiderwort flowers

Stephen as "of a good size and delicious", but do have the drawback of containing inulin. Perhaps best of all, the young flower spikes can be eaten. One Swiss website recommends pickling them.

Asian rapunzels have been moved to the genus *Asyneuma*. *A. japonicum*, with a native range up into the Russian Far East, has been recorded as being foraged in Japan.[46]

Spiderworts
Tradescantia species

COMMELINACEAE

I've always liked the spiderworts, better known by their Latin name of *Tradescantia* in British gardens, for their rich blue flowers and ability to naturalise in grass. Finding that they were edible was a decided bonus. There are 75 species, scattered across the Americas from Canada to Argentina, but only four edible temperate ones that I know of: *T. ohiensis*, the Ohio spiderwort or bluejacket; *T. occidentalis*, the western spiderwort (which looks identical to its Ohioan cousin to me); *T. virginiana*, the Virginia spiderwort; and *T. subaspera*, the zigzag spiderwort. I grow the first three but haven't been able to source the last yet, which is a pity as it is the most shade-tolerant. Garden spiderworts are hybrids, usually of the Andersoniana group.

A handful of spiderwort 'leeks'

Spiderworts grow in a slowly expanding clump, throwing up a dense, messy thicket of leek-like shoots. These can be harvested whole and cooked while young. Once they start to flower they become tougher and more fibrous and are not worth eating. Young leaves can be used as a pot herb or in salads and the flowers are edible too. I cannot find any information on the edibility of the hybrids but I have cautiously tried some without any ill effects. I suspect that they are all edible. When harvesting the shoots, the broken shoot exudes a clear sticky sap. I assume this is the source of another common name for a spiderwort: cow slobber!

The tender tips of orpine shoots

Orpine
Sedum telephium

CRASSULACEAE

Orpine is closely related to the popular butterfly attractor, ice plant (*Sedum spectabile*), but is a lot more shade-tolerant. It grows in a mass of shoots that remain tender at the tips for a long period. I harvest the tips for stir-fries and have known people who like the young leaves in a salad, although I find them too bitter. I have also read that the roots, which are large and fleshy, are edible, but sadly when I tried them I found them ferociously astringent, even boiled in a couple of changes of water.

Orpine grows in a tight clump, but tends to spread itself around the garden by seed (the flowers are far too pretty to remove) and perhaps by fragments of stem, which root easily. Despite this it is easy to control by pulling up unwanted plants.

Popular with pollinators – an orpine flower

Talet show: aerial beans, underground beans and leaf

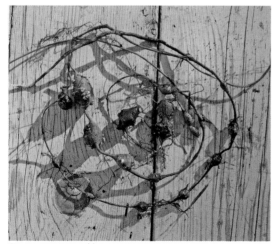
Hopniss tubers

Grow your own necklace

Talet
Amphicarpaea bracteata

FABACEAE

Talet, also known as hog peanut, has a fascinating lifecycle. As a member of the bean family it bears small seeds in pods in the normal way, but it also produces some larger seeds underground, like little tubers. Both kinds are edible, the aerial seeds cooked and the subterranean ones raw, although both are too fiddly to harvest to be a major crop. It grows wild as far north as Nova Scotia, but my plants (from Nova Scotian seed) only just cling on.

Hopniss
Apios americana

FABACEAE

Hopniss is another tuber-making, climbing, North American bean relative that mostly prefers to grow a bit further south than my garden – but unlike talet I do manage to get a worthwhile crop from it. It produces long strings of egg-sized tubers, decreasing from hen's egg at the start of the string to pigeon egg by the end. The strings radiate out in all directions from the plant and can be a challenge to trace. My strategy for growing it is to eat the larger tubers and pot up the smaller ones indoors. It will generally survive outside over winter, but doesn't come into growth until about June. Starting it off indoors and planting it out gives it a head start and gets a bigger crop. It will grow in full sun or partial shade. In the wild it twines up other vegetation, but since harvesting is quite disruptive to the roots of other plants and it can't take too much shade, I prefer to provide a few bamboo poles.

The taste of the tubers, which must be cooked to destroy protease inhibitors found in the raw form, is very good. It has a nutritional profile somewhere between a bean and a potato,[47] which seems appropriate. Sam Thayer gives a recipe for 'refried' hopniss as a filling for tacos and burritos. I mostly use mine in winter stews. You will notice a

Hopniss taking advantage of a leek flower to climb up

difference between the tuber you planted or left to grow and the new ones that grow on the strings. The old tuber is not used up like a potato is, but grows larger and gnarlier. It is edible, if a little rougher and in need of peeling. Tubers store well in a fridge or cold store if protected from drying out. They can also be preserved by drying, although I never have enough to make that worthwhile.

Aardaker
Lathyrus tuberosus

FABACEAE

Aardaker, also known as the Fyfield pea or Dutch mice, is a small, scrambling vetch with pretty pink flowers. It produces small underground tubers whose conical shape really does make them look a little like mice. Both taste and texture are pleasant and they go well in winter stews. Their size makes them rather fiddly to harvest and prepare, but a collective breeding programme to increase the size is already showing impressive results, with one cultivar yielding hand-sized tubers.

The shoots taste rather like pea shoots. A paper from Turkey describes them as being locally consumed as food in the Upper Çoruh Valley, and measures their nutrient profile. It turns out that they are the most nutritious of the eight species tested, as well as scoring better than many cultivated greens.[48]

Aardaker has deep roots, making it quite drought-tolerant. The roots spread through the soil, so an individual plant will form a clump, with the top growth scrambling up and over neighbouring plants. Because the leaves are quite fine, I haven't noticed this inconveniencing them much.

Aardaker root selected for size

Aardaker root selected for short runners for convenient harvest

Aardaker flower

Golden oregano as an 'edimental' in a park planter, Sunnybank Park, Aberdeen

Hoverfly *Eristalis pertinax* on oregano flowers

Oregano
Origanum vulgare

LAMIACEAE

Oregano is a well-known herb of Italian cooking, but there are also Scottish cultivars like 'Tomintoul' which do better in our climate. Indeed, it can become rather invasive in a garden and it is best not to let it seed. It likes full sun but will also tolerate a fair degree of shade. It is very easy to dry and I usually have a bunch hanging up in my kitchen. If you dry it when it is in flower it doesn't taste any better, but it does look spectacular once the dried leaves and flowers are crumbled into a jar. Bees also love the flowers. It grows to about 30cm (1ft) and is rather straggly.

Lilies
Lilium species

LILIACEAE

Many plants have the word 'lily' in their names, but the 'true' lilies are those in the genus *Lilium*. There are lots of them, and I have only just scratched the surface of their possibilities in my garden. China is the centre of their cultivation and the Flora of China, available online, includes many edible kinds.

Those with a northern distribution, forest edge habitat and edible use are listed in the table below. North America also has its species, including meadow lily (*L. canadense*), Columbia lily (*L. columbianum*), wood lily (*L. philadelphicum*) and swamp lily (*L. superbum*). Central Europe has *L. bulbiferum*, the fire lily. Many more are listed in compendia like Tanaka's *Cyclopaedia of Edible Plants of the World* and *Kunkel's Plants for Human Consumption*.

Generally it is the starchy bulb that is used, although some sources refer to parts such as the young leaves (*L. concolor*) or an oil from the flowers

Cluster of lily bulbs with loose scales

used as a spice (*L. brownii* var. *viridulum*). The two that I grow, the tiger (*lancifolium*) and Turk's cap lilies, both have mild, starchy bulbs, but the Columbia and wood lilies have a bitter, peppery taste and are used in interior British Columbia like garlic or chilli to flavour dishes. Nancy Turner describes it as an acquired but popular taste.[49]

All lilies can be propagated by lifting and dividing the bulbs. If you want to bulk up quickly the bulbs can also be split into individual scales, each of which will produce a little plant. A few plants, such as *L. lancifolium* 'Splendens' and *L. bulbiferum*, help the gardener out considerably by producing bulbils all along the length of their stems. These can be potted up and grown into mature plants. Growing from seed is trickier. Some species, including *L. martagon*, simply put down a root and create a tiny bulb in their first year, with no visible activity above ground at all. It is important not to throw these out as failures! All seed-grown lilies need nursing for at least a couple of years before planting out.

Lily growers recommend growing them with their 'roots in the shade and tops in the sun'. This happens quite naturally on the forest edge where they are surrounded by other vegetation (see page 16), but it can also be achieved by placing rocks to shade the roots.

Lilium lancifolium 'Splendens'

Lilium martagon

Bulbil on *L. lancifolium* 'Splendens'

EDIBLE FOREST EDGE LILIES MENTIONED IN THE FLORA OF CHINA

Species name	Chinese name	English name	Habitat	Notes
brownii	野百合 yě bǎihé	wild lily	forest edge	
brownii var. *viridulum*	百合 bǎihé	common lily	cultivated	Flowers contain aroma oil, can be used as spices
concolor	渥丹 wò dān	morning star lily	grassland and scrub	Young leaves cooked (Kunkel)
pennsylvanicum (*dauricum*)	毛百合 máo bǎihé	Siberian lily	forest edge	
davidii	川百合 chuān bǎihé	stream lily	forest edge	Bulb containing starch, excellent quality, high yield cultivation, edible
distichum	东北百合 dōng běi bǎihé	northeast lily	forest or forest edge	
lancifolium (*tigrinum*)	卷丹 juǎn dān	tiger lily	grass and scrub	The bulbs are rich in starch. Flowers containing aromatic oils can be used as spices.
martagon var. *pilosiusculum*	新疆百合 Xīnjiāng bǎihé	Turk's cap lily Xinjiang lily	mountain shade or under bushes	
pumilum	山丹 shān dān	coral lily mountain lily	hillside meadow or forest edge	
rosthornii	南川百合 nán chuān bǎihé	southern stream lily	perhaps not hardy	Use mostly medicinal
speciosum var. *gloriosoides*	药百合 yào bǎihé	Oriental lily medicine lily	forests and grassy slopes	Very showy, use mostly medicinal

Mallows
Malva species

MALVACEAE

If you like both your food and your vegetable garden to look beautiful, then the mallows are the plants for you. I grow three: musk mallow (*M. moschata*), Chinese mallow (*M. verticillata*) and common mallow (*M. sylvestris*). Musk mallow is the prettiest – even the names are evocative. Its best feature is its flowers, which come either in fairy white or candy pink, are produced throughout the growing season and have a melt-in-the-mouth texture when used in salads. Their one drawback is that they don't keep for very long once picked and will go rather yucky left in a salad bowl overnight. Their leaves are also useful and – yet again – very attractive, with deeply-cut palmate lobes. They taste fine and

look great in a salad but unfortunately I'm not very keen on the texture and prefer them cooked as a pot herb. They are a useful plant for this as they carry on producing leaves throughout the winter. They like a sunny spot and are robust enough to hold their own in grass.

Common mallow has the same uses as musk mallow, plus its seed heads, known as 'cheeses' for their wheel-like shape, are also edible. It is more of a woodland plant than *moschata*, and tolerates deeper shade. The whole genus, plus close relatives, is susceptible to *Puccinia* rust, which causes orange flecks to appear on the leaves as if they have indeed gone rusty. Common mallow suffers the most, but some cultivars of mallow, such as 'Zebrina', have been bred for a degree of resistance. These resistant strains aren't the hardiest, so it is worth crossing them with a locally-collected one.

Common mallow flower and cheeses

Chinese mallow, *Malva verticillata* 'Crispa'

Musk mallow is reliably perennial, while common mallow hovers somewhere between biennial and short-lived perennial. Chinese mallow is straight-forwardly annual, but maintains itself by self-seeding. Otherwise it is similar. A variety called 'Crispa' has attractive, crinkled leaves.

The people who appreciate mallows most seem to be the Moroccans, and all the best mallow recipes come from there. It's used to thicken a soup with lentils, beans and spices called *harira*, which is used to break the fast during Ramadan. All members of the mallow family have this thick-ening effect, with the extreme example of okra for gumbo. For a Scottish version of *harira*, the forager Robin Harford suggests adding wild celery.[50]

Musk mallow with visitor

Columbines
Aquilegia species

RANUNCULACEAE

Columbines appear on many edible plant lists. They are undoubtedly attractive and hardy plants, but I find their edibility rather marginal. The flowers are the least bitter part and certainly look pretty in a salad. Some people extol the early leaves for salads but I find them far too bitter. They are quite weedy: I have been trying to eliminate them from my garden for about 10 years and they still come up from seed.

Bride's feathers
Aruncus dioicus

ROSACEAE

Bride's feathers is a large herbaceous perennial native to damp woodlands across northern Europe, Asia and North America. It is very ornamental plant, with graceful foliage and great plumes of tiny white flowers that won it the Royal Horticultural Society's Order of Merit. The main interest for the home gar-dener comes earlier than this: they have one of the few edible shoots in the rose family.

Bride's feathers have a place in foraging traditions in a number of parts of the world. In Italy they are used in *pistic* leaf soup or boiled briefly in herb-infused water.[51] In Korea they are called *samnamul* and are used as mountain vegetables. The Slow Food Foundation mentions that on Ulleung Island they are called 'meat plant' (*gogyuinamul*) as they are considered to have both the taste and feel of meat.[52] To me the most pronounced flavour is a 'rosy' one which can also be found in raspberry shoots.

Strawberries and alpine strawberries
Fragaria × *ananassa* and *vesca*

ROSACEAE

While true wild strawberries will manage in deep shade, two other members of the genus, both with a history of cultivation, are better suited for forest edge or sunny beds. Alpine strawberry is not a separate species, but a form of the European wild strawberry (*F. vesca*) that has lost the ability to make runners. This single trait makes it a considerably different beast in several ways. Firstly, it makes it far easier to breed improved varieties,

Bride's feathers in flower

since individual plants stay as identifiable individuals rather than becoming a hopelessly tangled mixture. Over time, alpine strawberry cultivars have been selected for traits such as larger berries or white fruit that is left alone by the birds. Because they don't put their resources into vegetative reproduction, all their energy can be channelled into fruiting. The alpine strawberry season usually starts around May or June and lasts through to the first frosts. They are also much better behaved in the garden, forming neat little mounds about a

Hybrid strawberries in flower in the ground layer

foot in diameter that are very useful for lining paths that you can browse your way along.

By contrast, the cultivated or garden strawberry, *F. x ananassa*, arose as a hybrid between two New World strawberries: *F. virginiana* from eastern

Hybrid strawberry fruit

Alpine strawberries

Strawberries on a wood violet plant? Fruit from a plant hidden in the ground layer.

North America and *F. chiloensis*, which had been cultivated by Mapuche and Huilliche gardeners in what is now Chile. Garden strawberries can be grown in the familiar way in annual beds, with well fed and watered plants in beds that are dug in and rotated after three years to avoid the build up of viral diseases. They will also grow on a more informal basis like giant wild strawberries in forest edge or sunny beds, self-propagating by runners and popping up here and there between more static plants with unexpected offerings of berries.

Hybrid strawberries have largely replaced the older, *vesca*-derived cultivars due to their size and yield, but I find that they have complementary qualities that make it worth growing both. A patch of hybrids will give you a large glut of high summer berries, suitable for milkshakes, jam, fruit leather or strawberries and cream. Alpines give a longer season of smaller harvests, suitable for mixing with other fruit or simply savouring their distinctive vanilla-tinged flavour as you wander around the garden. They also dry easily on a sunny windowsill, scenting the room as they do.

Although they have lost the option of multiplying by runners, alpine strawberries are simple to propagate. They grow well from seed and will self-seed around the garden thanks to the birds. For clonal propagation to increase a superior variety, clumps can be lifted and divided in winter. Seeds are harder to remove from garden strawberries, with their larger fruit – I shave off and dry strips from the fruit, then scrape the seeds off the strips and sow in spring.

Miñe-miñe, Falkland Island raspberry
Rubus (Comaropsis) geoides

ROSACEAE

Miñe-miñe is a raspberry that thinks it's a strawberry! Chilean sources[53] describe the taste as being close to strawberry and indeed its names there include *frambuesa silvestre* and *frutilla silvestre*

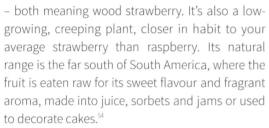

Miñe-miñe mine. Falkland Island raspberry seedling.

Rubus arcticus in Oulu, Finland

– both meaning wood strawberry. It's also a low-growing, creeping plant, closer in habit to your average strawberry than raspberry. Its natural range is the far south of South America, where the fruit is eaten raw for its sweet flavour and fragrant aroma, made into juice, sorbets and jams or used to decorate cakes.[54]

The *Flora Chilena*[55] describes it as growing best in 20-40% shade and tolerating temperatures down to at least -15°C (5°F). The plants in my garden, grown from seed, have so far been very slow growing and haven't fruited yet, so I can't yet say if its flavour lives up to Spanish-language reports.

Arctic raspberry
Rubus (Cylactis) arcticus

ROSACEAE

At the other end of the world there is another low-growing raspberry, *Rubus arcticus*, variously known as the arctic raspberry, bramble or plumboy, or the nangoonberry. It grows right around the Arctic Circle, in Alaska, Canada, northern Europe and northern Asia, but also comes as far south as

Colorado. Interestingly, it seems to be a lost Scottish native, described in the Online Atlas of the British and Irish Flora as having grown on Scottish mountains, habitat unknown.[56]

North American sources usually describe nangoonberry as growing on damp, peaty soils, but in Finland it grows in open, damp woods of birch, pine or fir, and in Estonia it is found in bogs, wet meadows, ditches and open spruce-hardwood woods,[57] making it sound like a good candidate for a woodland edge position in a forest garden. Despite its presence on damp soils in the wild, Kore Wild Fruit nursery, one of the few suppliers in the UK, say that good drainage is necessary for them to thrive.[58] It probably goes without saying that it is very cold-tolerant.

It certainly sounds worth growing, as all sources, from Linnaeus[59] onwards, mention its excellent flavour, with hints of pineapple. The Russian name *knyazhenika* means the 'berry of princes'.[60] Plants For A Future describes it as fruiting poorly in Cornwall, but it was one of the most productive of a wide range of northern berry crops evaluated by the Orkney College's Agronomy Institute in 2006.[61]

Chopsuey greens

11

Sun-loving Crops

Our normal temperate agriculture is so thoroughly based on plants of open ground and full sun that we forget that low-growing plants in full sun are, ecologically speaking, the exception. Any piece of ground with plants growing on it will tend, over time, to become dominated by trees. The trees trump the competition simply by growing taller and being the first to intercept the sunlight. Other plants are relegated to a life powered by the scraps of sunlight that the trees don't use. The only exceptions to this rule are where some constraint prevents trees from getting established, leaving the low-growing plants to bask in the sun in peace.

There are many possible constraints. Extremes of temperature and humidity both make life hard for trees. Very thin soils don't give enough purchase for roots so tall plants blow over. Very dynamic environments, like some streamsides and riversides, or shifting dune systems, change too rapidly for long-lived trees. The exposure and salt spray of the coast take their toll. In some parts of the world, trees like acacias or giant redwood have fireproofing that allows them to ride out fires (giant redwoods even start them, by conducting lightning strikes through channels in their bark, to clear out the competition), but in other parts of the world fire restarts the race to the sky and gives low-growing plants their chance.

The upshot of this is that sunny environments are far more varied than wooded ones. The plants that grow there win their place by being adapted to conditions that clear out the competition for them. It's not that the plants wouldn't be happy growing in better conditions, but the adaptations are usually costly, meaning that the plants lose out in open competition. In the tough conditions that they are adapted to, their reward is an uninterrupted stream of life-giving sunlight. To grow them in a garden we have to employ one of two strategies: either we can recreate the conditions that favour them or we ourselves have to take on the role of the ecological constraint ourselves, by weeding out competitor plants.

This chapter is therefore broken down by habitat, grouping the plants by the conditions that favour them best. Use this as a guide to where to plant them, grouping them together in niches, but don't be too constrained by it. Sun-loving plants are opportunists and there is a lot of overlap between open habitats, with the same species sometimes appearing on both sea cliffs and mountain ledges. I have put each plant or group of plants into the habitat where I feel it sits most naturally, but that does not mean at all that they will only be found in that habitat.

The coast

Many sun-loving plants are found on the coast, where exposure, salt spray and shifting sediments combine to make life hard for shading trees. Leaves are often reduced, waxy or silvery, to protect against sun and salt, and a spreading habit helps plants to hold their own in shifting sands. A sunny-edge bed mulched with seaweed provides perfect conditions for many of them. The more leafy species can also do well in a forest edge situation, where we have already met dittander (in 'horse-radish flavours'), sea beet and Scots lovage (in the licorice-roots, Chapter 10). The native alliums, sand leek and crow garlic, (in 'Sun-loving alliums', p.240) grow in mostly coastal locations and some perennial leeks (Chapter 10) have naturalised there too. Saltbush has been described in p.143.

Garden asparagus
Asparagus officinalis

ASPARAGACEAE

The traditional growing conditions of asparagus, on rigorously weeded and watered mounds with lots of compost, owe a lot to its origins on sandy or silty river banks and seashores. They make it a work-intensive crop, not fitting the forest garden ethos very well despite its perennial nature. You could experiment, like Andy Williams at Cairn of Dunn croft, with planting it on the banks of drainage channels. It is dioecious, and traditionally male plants are planted to avoid them expending energy in producing seeds.

Sea kale
Crambe maritima

BRASSICACEAE

Sea kale is a native of rocky beaches, endangered in the wild due to over-harvesting in Victorian times. Rather tough and strongly flavoured in its natural form, the leaves are usually blanched for eating, producing shoots described as a cross between cabbage and asparagus. Like many sea-shore plants, it is not very tolerant of competition.

Bladder campion
Silene vulgaris

CARYOPHYLLACEAE

It takes some work to collect enough of the fine shoot-tips of bladder campion for a meal, but it is worth it. As long ago as 1783, Charles Bryant declared, in the *Flora Dietetica: Or, History of Esculent Plants*, "Our kitchen gardens scarcely furnish a

Sea kale in flower

Sea campion flower

better flavoured sallad [vegetable] than the young, tender shoots of this plant, when boiled.'" It is a low-growing, mat-forming plant, easily overwhelmed by more vigorous neighbours, but will thrive on the edge of a bank and can naturalise in grass. The closely related sea campion (*S. uniflora*) can be used in the same way.

Storksbill
Erodium cicutarium

GERANIACEAE

I have always felt sure that the Geranium family must have some members suitable for a home garden, as so many have strong, interesting scents and so many grow as vigorous wildflowers in Scotland, many of them ending up as garden weeds. Unfortunately, it always seems that those that taste good don't grow well and those that grow well don't taste good. I had almost given up when I came across a mention in *Wild Food*, by Ray Mears and Gordon Hillman, of *Erodium cicutarium*, the common storksbill, a low, straggling annual or biennial plant, adapted to dunes and dry grassland. Its leaves are mild enough for salads and the shoots are very pleasant steamed. Although not long-lived, it will maintain itself by self-seeding.

Silverweed
Potentilla anserina

ROSACEAE

Silverweed is another seaside plant that uses a spreading root system to hold its own in shifting sediments. It is often seen on roadsides, where the salty conditions suit it well. The roots are both a blessing and a curse to the home gardener. They bear tuber-like thickenings which are starchy and delicious, and there are reports of it having been grown, very productively, in the Scottish Highlands in lazy beds (mounds mulched regularly with seaweed).

Silverweed in flower

Storksbill flowers

Silverweed roots

233

The downside is that they are extremely invasive, and need careful containment if they aren't to take over your whole garden. An advantage of growing them in pots is that the tubers mainly form at the bottom of the pot, and can be harvested easily by turning out the contents in winter.

Streamsides

Streams and rivers create special environments. Rather like coastlines, they can be dynamic environments, with floods and shifting sediments creating a niche for sun-loving plants – but without the exposure and salt stress of the seaside. The humid environment they create is perfect for delicate edible plants like the golden saxifrages (*Chrysosplenium alternifolium* and *oppositifolium*).

Some plants like to grow actually in the stream. Watercress (*Nasturtium officinale*) will grow in any wet soil but only survives the winter where it is protected by immersion in flowing water. Wasabi (*Eutrema japonica*) needs cold, clean, shallow running water to flourish. Unless you are lucky enough to actually have a stream in your garden, these conditions are quite hard to provide in a home garden. Bankside plants are far easier, requiring only some watering when the weather is dry. Daylilies and hostas (Chapter 10) do well in these conditions, as do a few alliums such as mouse garlic (*A. angulosum*) and shortstyle onion (*A. brevistylum*). See p.261 for plants of still water, marsh, bog and mire.

The Mediterranean

Mediterranean plants often look similar to seashore ones, with narrow, silvery foliage to ward off the sun. The Mediterranean climate is a mirror image of the cool temperate one, with the dormant season during the punishing heat of summer and the growing season in the mild winters, which can lead to Mediterranean plants putting on growth at unexpected times of year. Their big enemy is winter wet, so the most important thing is good drainage, which can be achieved by planting in a raised position or digging gravel into the soil. With this, they will tolerate surprisingly cold conditions, but they will still do best in the warmest, sunniest spot you can find for them.

Mediterranean climates are not only found around the Mediterranean Sea, but also in California, central Chile, the Western Cape, and south and southwest Australia. If wine is produced in a region, it's a fair bet you'll find Mediterranean climate plants there.

Mediterranean alliums
Allium species

AMARYLLIDACEAE

A number of alliums call the Mediterranean region home and will flourish in a Mediterranean bed. They are mostly at the garlicky end of the allium spectrum. Golden garlic (*A. moly*) and small yellow onion (*A. flavum*) both provide leaves and unusual yellow flowers for salads. Golden garlic also produces worthwhile garlic-flavoured cloves, in loose clusters rather than in the neatly wrapped bulb of cultivated garlic. Keeled garlic (*A. carinatum*) is closely related and similar to *A. flavum*, but with a more standard purple colour. The subspecies *pulchellum*, widely sold as an ornamental, is less likely to be invasive than other forms as it lacks bulbils. Rosy garlic (*A. roseum*) grows well in Aberdeen and extends the colour range to pink. I don't know how daffodil garlic (*A. neapolitanum*) got its common name, since it produces a loose cluster of ice-white flowers rather than large yellow ones. It is one of the most highly rated by Ken Fern for flavour, but unfortunately it survives rather than thrives in Aberdeen and I rarely dare eat many of its leaves for fear of killing it.

Fennel, with last year's stems and spring growth

Fennel
Foeniculum vulgare

APIACEAE

Fennel is a tall, graceful, herbaceous perennial with feathery leaves. Its stems die in autumn but usually stand until spring, giving structure to a herbaceous bed in winter. A variety called bronze fennel is especially striking, with dark purple foliage.

All parts have an aniseed flavour that is traditionally paired with fish. While the mature foliage is normally used as a herb, the fresh growth is, as usual, milder and tenderer: this includes the basal leaf shoots, the flower shoots and the stem leaf shoots. A selection called Florence fennel has swollen leaf bases but needs to be treated as an annual with lots of warmth, feeding and water. I have tried the roots, which are thick and fleshy, but they are disappointingly bland for a member

of the carrot family. The seeds, on the other hand, are especially good. Picked green and dried they make an excellent spice and are chewed after meals in some countries to freshen the breath and (allegedly) reduce the chances of gas.

Biscuitroots
Lomatium species

APIACEAE

If there was ever a promising name for a group of plants I think 'biscuitroots' has to be it. My hopes for the genus rose even higher when I read that one species (*L. dissectum*) is sometimes called 'chocolate tips', but sadly that turns out to be for colour, not taste. Nancy Turner lists eight different species of *Lomatium* as having been used by First Peoples of inland British Columbia.[2] Seeds are not easy to get hold of in Europe and I am only at the early stages of experimenting with three of them.

The biscuitroots are all herbaceous perennials growing in dryish environments in the west of North America. Both shoots and roots have been used. As well as biscuitroots many of them are known as desert parsley or wild celery, reflecting their affinity with relatives in the Apiaceae. Indian celery (*L. nudicaule*) sounds like the best for shoots. Young sprouts, leaf stalks and leaves were all popular, and boiled Indian celery was considered a special treat for children. The greens are still gathered and preserved by freezing or canning. Along with two other species, *triternatum* and *ambiguum*, the flowers, leaves, stems and seeds of Indian celery are used variously as herb, tobacco and tea.

Chocolate tips was used for both shoots and roots. The roots have a bitter taste that is enjoyed by some, giving the plant another name: bitter-root. The roots can be dried for storage. *L. macrocarpum*, one of the species known as desert parsley, was also used for its roots. They were dug in spring and have a peppery, celery-like taste. Finally, groups in British Columbia traded for three species of biscuitroot roots, *canbyi*, *cous* and *farinosum*. These grow somewhat further south and inland and may be less well adapted to cool temperate conditions. All useable *Lomatium* roots can be dried for later use.

Wild asparagus
Asparagus acutifolius

ASPARAGACEAE

Wild asparagus is related to garden asparagus, but with finer, spinier stems that suit its name. It wins no prizes for productivity, but has an intense asparagus-like flavour that makes its cultivation worthwhile. I mistakenly grew mine in my seashore bed for many years, thinking it was simply the wild ancestor of garden asparagus, but in fact it is a more Mediterranean species, suited to a sunny position on freely draining soils.

Bath asparagus, with black rapunzel (*Phyteuma nigra*) in the background

Bath asparagus
Ornithogalum pyrenaicum

ASPARAGACEAE

Bath asparagus is another plant in the asparagus family, and its flowering stem is eaten in a similar way. Stephen Barstow says that he lost his over winter on his clay soil, while mine manage on my sandy soil, so it is probably a typical Mediterranean plant in this sense, needing to avoid winter wet in the north. It grows in woods and scrub in its native climate, but mine only grow very slowly in the shady position I have given them. In cooler climates a more open spot would probably be beneficial.

King's spear
Asphodeline lutea

ASPHODELACEAE

King's spear does most of its growing in winter, producing abundant bundles of silvery leaves that can be trimmed and eaten a little like leeks.

Individual king's spear flower

King's spear in suitably exotic surroundings in the Summer Garden in Aberdeen's Duthie Park

Immature flower shoot of king's spear

In summer it produces a spectacular flower spike that presents the home gardener with something of a dilemma. The flowers, which individually live for only a day or two but are borne on the spike over a long period, are very pretty and have a lovely sweet flavour. However, to my mind the part that tastes best is the entire flower spike, picked when young and steamed – meaning that you miss out on all the flowers. The roots are also said to be edible but I find them fiddly and tasteless.

Globe artichoke
Cynara cardunculus

ASTERACEAE

All the Mediterranean plants look exotic, but none quite so much as this species. *Cynara cardunculus* exists in two forms: the globe artichoke, bred for its edible flower heads; and the cardoon, bred for

the thick midribs of its leaves, which are usually blanched for eating. Received wisdom has it that cardoons are hardier than artichokes so this was the form that I tried first, but I have come to the opposite conclusion – that artichokes are the only form worth growing in cooler climates.

Both forms grow more slowly in cooler areas than in their native range. This is a problem for cardoons as their culture is based on treating them as an annual, and I haven't been able to develop a perennial equivalent that is worth the effort. Globe artichokes, on the other hand, thrive as perennials, becoming more productive year on year. Their comparative frost-tenderness also seems to be a myth.

I once came up with the idea of a weight-control diet called the Fiddly Food Diet. Its premise was simple: you can eat as much as you like, but only of fiddly foods that take a lot of work to get into,

The striking silvery form of a globe artichoke plant

Too late for eating, but perfect to show that an artichoke is basically just a ginormous thistle

Cardoons on sale in a grocer's in the Basque Country

like beech nuts and pomegranates. Globe artichokes would definitely be on the menu. They are a theatrical food rather than a filling one. The edible part is the immature flower head, but only a little bit of tender flesh at the base of each scale and the 'heart' underneath the 'choke' of the inedible flowers. Steamed for 20-30 minutes until the scales come away easily, then dipped in olive oil and lemon juice, they are well worth the effort.

Give artichokes typical, well-drained Mediterranean growing conditions. They will try to grow during winter, and new growth can be cut back by frosts, but they are surprisingly hardy and will grow right through a mild winter here in Aberdeen. They have an extensive fleshy root system that means they can survive being frosted and come back as good as new once they are well established, but give them some frost protection in their first winter or two. In continental climates with guaranteed long, freezing winters, it is best to give them a thick mulch every winter. Cold-hardy cultivars include 'Green Globe' and 'Northern Star'.

Perennial wall rocket
Diplotaxis tenuifolia

BRASSICACEA

The seeds and leaves of perennial wall rocket are both so regularly sold as 'wild rocket' that it can be difficult to separate from its annual cousin, *Eruca versicaria* (Chapter 10), but there are significant differences between the two plants. They taste quite similar, but wall rocket has a hint of cabbage and more of a kick. Ecologically, they are much more different. While *Eruca* is an annual that grows well with a bit of shade, wall rocket is a perennial adapted to growing in dry, rocky conditions with plenty of sun. It has a thick, fleshy root system that allows it to withstand some drought and resprout quickly afterwards.

The leaves of wall rocket get stronger as they get older, but there is a continual production of fresh leaves, so I don't find this matters much. You can

Perennial wall rocket thriving on a well-drained mound

cut the plant down to encourage a bigger flush of new leaves. If you have lots of older leaves, you can blanch them in boiling water for about 30 seconds which reduces the heat a bit. Cooking them for longer than this isn't recommended as they lose their flavour entirely. My favourite uses are in salads, sprinkled on pizzas once they come out of the oven and as a topping for pasta sauces. You can also make soup with them in the same way as you make watercress soup.

Red valerian
Centranthus ruber

CAPRIFOLIACEAE

Red valerian is an attractive plant, often found growing on walls, related to honeysuckle. Its niche in the forest garden is supplying salad leaves in the winter and early spring when not much else is growing. They are bitter, but pleasantly so. Later in the year the bitterness increases and we must turn, like the butterflies, to appreciating the showy red

Red valerian leaves in winter

Red valerian flowers in summer

239

flowers. Its favoured spot is dry with an alkaline soil, but I have found it to be quite adaptable and capable of taking light shade as well.

Sage and rosemary
Salvia officinalis and *rosmarinus*

LAMIACEAE

Sage and rosemary not only share a line in 'Scarborough Fair': since a recent reorganisation they now share a genus, with rosemary having been renamed from *Rosmarinus officinalis* to *Salvia rosmarinus*. This seems appropriate, as they have a lot in common. Both are aromatic Mediterranean herbs that need no introduction in terms of use. Both dry easily if hung up in a well-ventilated place. Both can grow to be medium-sized shrubs in ideal conditions but often die, partly or completely, in a northern winter. The key to growing both is to give them a very well-drained soil and a sunny spot, as they can tolerate drought but not waterlogging, especially when it is cold. Fortunately, both root easily from cuttings during summer so you can take backup copies and keep them somewhere frost-free over winter in case the main plant doesn't make it. Plant these out as soon as possible to give them the best possible root development before the next winter comes.

Rosemary in a planter in Sunnybank Park, Aberdeen, showing that it is a very ornamental plant as well as a tasty one

Freely draining soils

Where Mediterranean plants need freely draining soils to help them avoid root rot in winter, others prefer them because they have adapted to growing on them in cooler climates. Steep slopes and sandy soils can both produce conditions where water doesn't hang around for long.

Sun-loving alliums
Allium species

AMARYLLIDACEAE

We have met *Allium* species in many places throughout the forest garden, but when gardening guides talk about 'the alliums', they usually mean these: *Allium* species adapted to free-draining soils in open sun. There is a bewildering number of them, and it is easy to succumb to collector's mania, but in reality you don't need to get them all, just a large enough variety to make sure that you have fresh oniony leaves at most times of year.

Alliums have a broad range of reproduction strategies, giving both a lot of options for propagating them and a plethora of edible parts. The bulb is too fiddly to bother with in most; the biggest exceptions are the cultivated alliums (p.247) and Persian garlic (*A. altissimum*), which has big, garlicky bulbs that can be used fresh or sliced and dried.

Many species don't form large bulbs at all but grow on a rhizome and make dense clumps of stems. Chives (*A. schoenoprasum*) is the best known of these 'rhizomatous onions', which are mostly used for their leaves, either picked individually or from the base as a bundle. Others include mouse garlic (*A. angulosum*), a native of wet meadows from Europe to Siberia that likes a moist but well-drained soil; blue chives (*A. nutans*), a very variable species from Siberia; broadleaf chives (*A. senescens*), which has a pleasing helical twist to its leaves; and Chinese chives (*A. ramosum*), from Mongolia and surrounding countries. In one species, *A. hookeri*, the rhizome itself is thick and fleshy enough to be a

Allium height competition: *Allium altissimum* (the very tall onion) is pipped at the post by a humble leek

A clump of *Allium altissimum* (Persian garlic) bulbs

food item itself, but unfortunately this species did not thrive when I tried it as it has quite a southerly distribution.

I became quite excited about another rhizomatous onion, the Pacific onion (*A. validum*), when I first heard about it, as another of its common names is swamp onion. An allium for wet ground? It is also quite productive, with substantial leaf bundles growing from a thick rhizome. Unfortunately, although it does grow in damp meadows in its natural habitat on the Pacific states of North America, it is intolerant of winter wet, so in a maritime climate the best conditions are probably on a well-draining soil with some watering in summer. Klamath-Siskiyou Native Seeds, which specialises in wildflower seeds from this region, recommends 60-90 days of cold stratification, so seed is probably best obtained and sown in autumn or early winter.[3]

Allium seedlings with their characteristic bent over look

The flower stem of alliums, called a scape, is often edible when young, as is the flower head. I make at least as much use of the flowers of chives as I do of the leaves. Allium flower heads all have multiple flowers, radiating out from a central point. The number of flowers and the length, shape and direction of the stems make for a range of styles, from a tight ball to a loose spray, but all can be turned into a shower of individual flowers by nipping out the central point.

A bumblebee enjoying a chive flower

The yellow flowers of *Allium moly*

The Norrland onion

Vivipary (live birth) in onions: a mix of flowers and plantlets on a hybrid chive

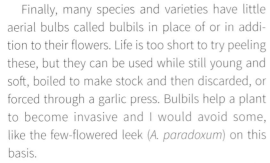

Offset bulbs on *Allium vineale*

Finally, many species and varieties have little aerial bulbs called bulbils in place of or in addition to their flowers. Life is too short to try peeling these, but they can be used while still young and soft, boiled to make stock and then discarded, or forced through a garlic press. Bulbils help a plant to become invasive and I would avoid some, like the few-flowered leek (*A. paradoxum*) on this basis.

There are two native sun-loving alliums in the UK, both quite closely related, both found mostly in sandy, coastal soils and both, unfortunately, not very productive. They are crow garlic (*A. vineale*) and sand leek (*A. scorodoprasum*). The latter is also sometimes known as rocambole, but shouldn't be confused with the cultivated garlic of the same name. Both have tiny bulbils and wiry scapes: sand leek also has a few small, garlic-flavoured bulbs. The only part I use in practice is the leaf bundles, like small but nicely-flavoured leeks, which appear as if from nowhere in thick clumps from shed bulbils in spring. In North America, meadow garlic (*A. canadense*) is quite similar.

Many alliums cross readily with other, related, species, and a garden with lots of them is likely to have new hybrids popping up on a regular basis.

This can lead to inferior plants, but also sometimes to especially vigorous ones. Stephen Barstow tells the story of the Norrland onion,[4] a cross between *nutans*, *senescens* and possibly a third, unidentified onion, that originally arose in Sweden. It is large, sterile and a vigorous grower, making it a promising candidate for the kitchen garden. I have at least one hybrid myself, grown from seed from Stephen's garden (he is an avid alliophile, so the potential for crosses is endless). It appears to be another *nutans-senescens* cross (they are closely related), and has rather endearing little viviparous onions growing in its seed head, complete with the characteristic twist in their tiny leaves. It doesn't seem to be as vigorous a grower as the Norrland onion, but I love it anyway.

Mugworts, wormwoods and sagebrushes
Artemisia species

ASTERACEAE

Artemisia is a large, diverse and widespread genus, mostly of perennials adapted to growing in dry places and fortified against browsing animals with strong-tasting essential oils and bitter chemicals like terpenoids and sesquiterpene lactones. As usual, we humans have commandeered these chemicals, and artemisias are used as medicines, herbs and flavourings for drinks. Artemisinin, from *A. annua*, is an important treatment for malaria. Absinthe wormwood (*A. absinthium*) is one of the main flavourings of the infamous spirit absinthe and also gives it name to the drink vermouth (via the German *wermud*). A number of members of the genus bear the common name 'mugwort', from their use as a flavouring in beer, similar to hops. Sheeba (*A. arborescens*) is used in the Middle East as a tea, usually with mint.

Currently I only use two members of the genus in my home garden, although I am convinced that the group would reward further exploration. *A. dracunculus* var. *sativa* is the French fine herbe

tarragon, commonly used for flavouring vinegar or pickled cucumbers, but also useful as a pot herb. *A. princeps* is known as *yomogi* in Japan and *ssuk* in Korea. Both cuisines combine it with rice to make rice cakes or dumplings, to which it adds a fresh, grassy flavour. In Korea it is also used for soup and kimchi. You can even get a *ssuk* latte! My favourite use for the shoots is as tempura; the underside of the leaves is covered with fine, silvery hairs which give it an unpleasant texture steamed or stir-fried but are perfect for holding on to batter.

These artemisias are herbaceous perennials, growing from a mass of underground runners. Clumps will expand, but not so quickly as to become invasive in my experience. They grow well in a dry, sunny spot like the base of a wall but also tolerate some shade for part of the day.

Calcareous soils

Another way to group plants is by pH. Some are adapted to growing on calcareous soils, where underlying rocks like limestone keep the pH high. As well as the species below, round-headed rampion (*Phyteuma orbiculare*, see the spiked rampions, Chapter 10), and the Mediterranean plants wild asparagus, king's spear and red valerian tolerate a high pH.

Spignel
Meum athamanticum

APIACEAE

Spignel is three plants in one! It has leaves that taste like dill, seeds that taste like curry and roots that taste like parsnip. In practice I don't use the roots much as they are small and require digging up the plant, but the leaves are a regular addition to salads, stews, stir-fries and leaf sauce, and the seeds can be dried and ground as a home-grown curry powder. It is a low-growing plant that makes a slowly-increasing clump and also self-seeds to a degree. A Scottish native, it is found in calcareous

Spignel

Salad burnet

grassland, but also seems very happy in a sunny, well-drained spot in my allotment.

Burnets
Sanguisorba species

ROSACEAE

Salad burnet (*S. minor*) is another native plant associated with grassy meadows and alkaline soils, although it seems quite tolerant of a range of conditions and I have also seen it growing in quite wet places in the wild. It is best thought of as a herb, with a unique flavour that is used in summer drinks, salads and sauces. It is often described as having a cucumber taste, which I completely fail to detect; it is possible that this only develops when the plant is on an alkaline soil. The young, expanding leaf shoots are milder and better textured than older ones. It is a low, straggling plant that will inevitably intermix with the foliage of others rather than making a neat individual.

Great burnet (*S. officinalis*) is similar but, not surprisingly, larger. It is used in traditional medicine in both Europe and China. Canadian burnet (*C. canadensis*) is a little different: it is native to North America, grows on wet soils, is larger again at over a metre tall and is best used cooked. The youngest leaf shoots are again the best part.

Prairie and steppe

Forest gives way to prairie or steppe in drier, continental regions. The original prairies were complex perennial ecosystems that have now been largely replaced with annual agriculture. There is a movement, analogous to forest gardening, to breed perennial food plants for prairie regions, but it is beyond the scope of this book and most of them would not grow well in a forest garden. A few plants of the prairie and steppe do, however, make the transition well. They are used to continental conditions, so they prefer a hot, sunny spot in summer. In winter they can tolerate cold but not waterlogging, so give them a well-drained, fertile soil.

Prairie onions
Allium species

AMARYLLIDACEAE

Alliums from the prairie include nodding onion (*A. cernuum*), the ubiquitous *A. canadense*, and two species that share the common name 'prairie onion': *A. drummondii* and *A. stellatum*. All are fairly typical alliums with edible leaves and flowers and small bulbs. My seed-grown nodding onions are pretty, but disappointingly small, so I'd advise getting

Nodding onion flowers in agreement

Colewort beginning to flower

one of the larger cultivars, such as 'Pink Giant', 'Major' or 'Hidcote', that are available.

Coleworts
Crambe species

BRASSICACEAE

Sea kale is the best known *Crambe*, but ecologically speaking it is a bit of an outlier in the genus. More typically, they grow on steppes. My favourite is *C. cordifolia*, known as colewort or flowering sea kale. It is a large, ornamental and hardy plant, with leaves growing up to a metre tall from a fleshy rhizome. They are quite substantial even if picked when young and tender. Boiled, they have a nice texture and a pleasant, mildly mustardy taste. They can also be blanched like sea kale, making them even milder. The immature flowers are even better: like broccoli with a hint of mustard. If you don't eat them, the flower head is truly spectacular: up to two metres tall and bearing a cloud of snowy white flowers. The seeds are large and, when young and green, can be picked and eaten like wasabi peas.

Then there are two species, confusingly called *C. tataria* and *C. tatarica*. For a long time I assumed that these were variant spellings of the same name, but the Plant List does list both as accepted names. The confusion is so widespread that it is

Colewort 'broccoli'

Colewort flowers

hard to know, if reading a description or buying seed, which one is really meant. Common names applied, fairly interchangeably, to the two species, include Tatar (or Tartar) breadplant, Tatar horse-radish and the Russian name *katran*. My plant was grown from seed bought as *tataria*, for what it's worth. It is smaller than colewort but can be used similarly. Some sources describe the thick, fleshy root as tasting sweet and/or turnip-like,* but those who have actually tried them tend to liken it to radish or horseradish.[5] The sweet roots may either be a vanished selection or just wishful thinking, copied from source to source. I would rather leave the root alone and feast on the flowers.

Milk vetches
Astragalus species

FABACEAE

Astragalus is a genus that I have only just begun to explore. *A. crassicarpus* is a prairie plant, known as 'ground plum' or 'buffalo bean' for its large, green, edible seed pods – a trait shared by many members of the genus. *A. canadensis*, or Canadian milk vetch, grows in a wider range of habitats and has successfully grown and overwintered in my garden. There are some records of the root being used, mostly medicinally, but mine has produced nothing that would tempt me to try it. *A. sinicus*, the Chinese milk vetch, is naturalised in Japan, growing in fields and dry paddies. The blog Shi-kigami says that, "The flowers have a delicate pea flavour, fresh and pleasantly sweet. The leaves are better cooked."[6] It may not be hardy in Scotland.

Low heath

Low heath species are adapted to surviving harsh conditions and having competitors controlled by those conditions. They might be worth growing in a home garden if you have a particular liking for one and/or you have an area of thin soil that really can't be improved, but by and large they are best left in the wild, as on improved soils they are unproductive and poor competitors for weeds.

Edible plants of low heath include small blueberries like *Vaccinium myrtilus* (blaeberry, billberry), *V. praestans* (Kamchatka bilberry or krasnyk), *V. myrtilloides* (Canadian blueberry), *V. scoparium* (grouseberry), *V. boreale* (northern blueberry), *V. uliginosum* (bog bilberry), *V. deliciosum* (Cascade blueberry) and *V. caespitosum* (dwarf blueberry). Other members of the heath family include bearberry (*Arctostaphylos uva-ursi*), cowberry or lingonberry (*Vaccinium vitis-idea*), crowberry (*Empetrum nigrum*), and the cranberries *V. macrocarpon* (American cranberry), *V. oxycoccos* (common cranberry), and *V. microcarpum* (small bog cranberry). Outside the heath family you might find thyme (*Thymus vulgaris*) and Scottish bluebell (*Campanula rotundifolia*, also known as harebell).

Agricultural crops and weeds

Another constraint on tree growth leading to sunny conditions has only emerged in the last ten thou-sand years – us! The advent of agriculture has created new niches, both for our crops (some of which have become so adapted to agriculture that their original habitat is unknown) and their weeds. A related niche is what is often known as 'waste ground': land cleared of trees and often nutrition-ally enriched by human activities, then abandoned.

Most agricultural crops are only suitable for annual beds outside the forest garden, but some, mostly those with large propagules that give them a fast start, grow well in patches in between the perennials in the sunny beds. A few typical weeds are also good enough in their own right to be worth maintaining as self-seeders. The crops in this section typically prefer fertile, well-watered soils.

---

* That is turnip (or neep) in the Scots sense: known elsewhere as swede or rutabaga. Common names are very complicated!

Epazote
Dysphania ambrosioides

AMARANTHACEAE

Epazote really shouldn't be able to grow in Scotland. Its native range is Central and South America and southern Mexico. Despite this, several plants have now passed several winters without any protection here in Aberdeen, so I'm ready to call it a success. It hasn't seeded yet and might need some winter protection to get it going fast enough to do so.

Epazote is a herb of Mexican cuisine, most often used to flavour black beans. It has a very complex flavour. Attempts at describing the taste and smell include comparisons to 'oregano, anise, fennel, or even tarragon, but stronger', 'turpentine or creosote', citrus, savory, and mint. I just describe it as delicious. As a bonus, it is famed for an ability to get rid of the fart-inducing properties of the beans. I haven't done enough controlled experiments yet to comment! In large quantities the leaves might make you feel sick, although due to the strength of the flavour I can't imagine that anyone would want to over-consume. The essential oil responsible is found in greater quantities in the flowers and seeds, which are best avoided.

Although now moved to a separate genus, epazote used to be in the genus *Chenopodium* and it is closely related to other goosefoots like Good King Henry and fat hen.

Cultivated onions
Allium species

AMARYLLIDACEAE

The familiar bulbing onion (*A. cepa*) does not grow well amongst perennials. It needs a very open spot and little competition. However, there are several other onions with a long history of cultivation and a liking for rich, agricultural soils that are themselves perennial and quite at home in the sunny beds of a home garden.

One of these is the welsh onion, which despite its name has nothing to do with the country of Wales. Its name is derived from an Old English word for 'foreign', pointing more accurately to its true origin in East Asia. DNA studies indicate that its wild ancestor is *A. altaicum*, from Siberia and Mongolia, and it has a long history of cultivation in China, Japan and Korea. It is closely related to bulb onion and tastes similar, but instead of forming a single, large bulb it divides repeatedly into lots of small ones. It is best thought of as a bunching spring onion. It can be used as such, especially in spring when the clump is regrowing after winter. Individual 'spring onions' are simply removed from the main clump, which carries on growing. At other times of year individual leaves can be harvested and the flower head, which is large and fleshy, can be used whole when young. *A. fistulosum* is also sometimes sold as 'Japanese bunching onion'. Japanese cultivars come in exciting red and white varieties but unfortunately don't seem to be so hardy in cool areas.

Two similar onions are *A. pskemense*, the Pskem River onion, which Stephen Barstow says is grown widely in botanic gardens in Scandinavia[7] and which might just be the wild ancestor of *A. cepa*; and *A.* x *proliferum*, the walking or Egyptian onion. Walking onion is a cross between *cepa* and *fistulosum*, which bears large bulbils in place of its

Epazote

Welsh onion clump separated

Welsh onion flowers unwrapping themselves

flowers. These plant themselves at a distance from the original bulb when the stem collapses in winter, allowing them to 'walk' around the garden. They can be used in similar ways to welsh onion but are more of a novelty than a really useful vegetable.

A number of other perennial onions are cultivated widely in East Asia. *A. chinense*, known as Chinese onion or rakkyo, unfortunately seems to prefer warmer conditions and has never flourished

Garlic chives

in my garden. *A. tuberosum*, or garlic chives, is hardier. It is very similar to Chinese chives (p.240). The common name gives a pretty good description of what to expect from it!

Finally, garlic (*A. sativum*) may be the oldest of all the cultivated onions, with its use going back at least to the Ancient Egyptians. For the classic bulb of garlic it is best grown in the annual beds, but I also have perennial clumps of garlic growing in 'rockery' situations in the perennial beds. These are used for what you might call spring garlic: spring-onion-like shoots with a garlic flavour, harvested either by cutting or by lifting and dividing the clump.

Skirret
Sium sisarum

APIACEAE

The roots of skirret are so sweet that they were a favourite of the Emperor Tiberius, a man who could presumably have pretty much whatever he wanted for his table. The roots are conveniently produced in bundles, allowing the gardener to

Skirret plants in spring

A skirret root bundle

Skirret flower shoots

simply lift a clump at the end of the year, pick off the best half and replant what is left. The immature flower shoots, which plants simply can't be discouraged from producing, are also very good, with a distinctly carroty flavour.

The roots only need to be cooked for a few minutes, and often burst open due to the amount of starch. Their main drawback is a somewhat fibrous core. If whole roots are cooked the flesh has to be stripped off the core by pulling it through the teeth. It is much more convenient to cut them into pieces, and my main use of the roots is to add sweetness and thickness to winter stews, where short lengths of the core become unnoticeable. The degree of fibrous core varies considerably from plant to plant, with some having a very noticeable one and others barely any. Having tracked a number of plants over several years, I haven't been able to find any very strong correlation between core and growing conditions or with individual plants from year to year. There does seem to be a certain degree of difference between plants though, so perhaps this trait could be bred out. It is an excellent candidate for plant breeding as it reproduces both sexually and vegetatively. The other drawback is that the roots are ridged all along their length and too thin to peel, making cleaning a bit of a task. I find an old toothbrush to be the perfect tool for the job.

Most of skirret's relatives are wetland plants, and it too prefers a well-watered, fertile soil. Stocks can be bulked up quickly by dividing clumps, which usually produce several separable 'crowns' from one every year. Plant crowns around 40cm (16in) apart. In areas where the soil is not frozen over winter the roots can be left in the ground until needed.

Pot marigold
Calendula officinalis

ASTERACEAE

Pot marigold is an annual plant closely related to the asters. Its bright orange or yellow petals cheer up both the garden and any food they are

Pot marigolds

sprinkled over. Its raw leaves can taste delicious but can also leave a scratchy feeling in the mouth. I prefer them cooked as a pot herb. The flower shoots can be stir-fried or used as a *namul*. I find that this species maintains itself by self-seeding without being invasive.

Chopsuey greens
Glebionis coronaria

ASTERACEAE

The Latin name of chopsuey greens used to be the much more sonorous *Chrysanthemum coronarium*.

A perfect 'chrysanthemum' flower on *Glebionis coronaria*. See also p.230.

Unfortunately it got caught up in the rearrangements of the Asteraceae and it now answers to *Glebionis*. It is similar to pot marigold in many ways, except that the flower buds are horribly bitter, so its leaves are the only part used. Like pot marigold it flowers quite late in the season, providing a summery feel to the garden right up until the frosts.

Sow thistles
Sonchus spp.

ASTERACEAE

I first learned about using sow thistles from Stephen Barstow, who details their use around the world in *Around the World in 80 Plants*.[8] They grow naturally in rocky crevices but find our cultivated fields and beds irresistible. Perhaps we should stop thinking of them as weeds, as they are very tasty vegetables in their own right. Both leaves and immature flower heads can be used: stir-fried, as a pot herb or as one of my favourite *namul*. If this sounds good to you, you might find it hard to buy seeds, but just wait a while and it will probably turn up in your garden anyway.

This is one case where perennial is not better. The perennial species, marsh sow thistle or *S. sonchifolius*, has less useable foliage and a far poorer texture to its flowers. The two annual species,

Smooth sow thistle in its favourite place, between pavement and wall

common sow thistle (*S. oleraceus*) and spiny sow thistle (*S. asper*), are better. The 'spines' of spiny sow thistle look fearsome but are in fact not sharp – but they are still better removed before eating. The common sow thistle has the best texture of all.

Like many relatives, sow thistles produce a milky sap as a defence when broken. This turns brown on exposure to air and stains anything it gets on, so I generally pick a handful and go and wash it at the allotment tap before putting them in anything. Leaves and stems both wilt very quickly, which I guess is why you will never find sow thistle greens in the shops. They are best picked straight before using.

Salsify's dandelion-like seed head

Salsifies
Tragopogon spp.

ASTERACEAE

Salsify (*T. porrifolius*) is best known as a root crop, with a delicate taste that gives it the alternative name of 'oyster root'. My favourite part, however, is the large, purple flowers that it bears in abundance in its second year. They can be used before or shortly after opening. Steamed and dressed with olive oil and lemon juice, they taste like artichoke without the fuss. The leaves are also edible: when they emerge in early spring they are tender and

A bouquet of salsify flowers ready for dinner

sweet and make a welcome addition to the range of salad leaves available; later on they become tougher but are useful as a pot herb for a while.

Salsify has a number of relatives that can be used in much the same ways: they are variously known as salsifies, goatsbeards or oysterplants. The meadow salsify (*T. pratensis*) is also known as Jack-go-to-bed-at-noon for its habit of folding its flowers away at midday. *T. dubius* is called yellow or western salsify. Despite its dubious name it is just as edible as the others. All of them will self-seed happily.

Alfalfa
Medicago sativa and *falcata*

FABACEAE

Alfalfa is grown widely as a forage and is becoming well known as a seed for sprouting, but it also makes a nice vegetable, producing masses of spring shoots that go well in stir-fries and green sauce. It is deep rooted and drought-tolerant, and doesn't spread from the roots like many related plants, but the top growth (of up to a metre) is rather straggly. It can be cut after the spring harvest for compost and to generate a new flush of growth. The two species of alfalfa are closely related and often crossed in fodder varieties.

Alfalfa

Alfalfa shoots

Broad beans
Vicia faba

FABACEAE

Peas, runner beans and broad beans are the only leguminous crops that will grow easily and reliably in a northern Scottish garden. I have experimented a lot with growing peas and runners in the hurly burly of the forest garden beds, reasoning that their ability to climb up sticks and other plants should be a help. It can be done, but in practice it is diffi-cult, and the yields aren't worth it. They do better in the more open conditions of the annual beds. Broad beans, on the other hand, are well suited to growing in little open patches in the home garden, as their large seeds and relatively tough leaves give them a fast start.

They are well worth making a bit of space for, as they are very useful plants. Their young leaves can be eaten in salads and the beans can be used fresh or dried. I usually soak dried beans for several days, during which time they mobilise stored starches into sugars and develop a rich, sweet taste. Cooked quickly in a pressure cooker and then used in winter stews or cooked in a little of their own juice with olive oil, garlic and lemon juice they are delicious (the same, incidentally, goes for

Broad beans come in lots of sizes, colours and patterns. I received this unusual marbled one in a swap.

Nodules on the roots of broad beans fix nitrogen directly from the air

252

runner beans, although for some reason in Britain we tend to stick to using the immature pods).

Broad beans are traditionally grown in double rows, with stakes at the ends of each row and string around the whole thing to stop the plants from falling over. This method transfers very well to the forest garden. The plants respond well to mulching with weeds laid alongside the rows.

Evening primrose
Oenothera biennis

ONAGRACEAE

Evening primrose is a biennial plant of disturbed ground, especially on light, sandy soils. Like many similar plants they have done well out of humans, both on our field edges and waste ground and grown deliberately in our vegetable gardens and flower beds. They were cultivated as a crop by a number of nations in eastern North America, before going worldwide on European contact.

In the first year the edible part is the root. The leaves of the basal rosette in the first summer and winter are strong and unpleasant, but young, fresh leaves in the next spring and on the growing stem are far milder and make a nice pot herb. The stems themselves can be peeled and eaten; they are sweet and make a good *namul*. Once plants start to flower, which they do in a wave from the bottom to top of a long flower spike, the flower buds, flowers and very young seed pods are all edible. The seeds themselves are produced in abundance and easy to collect. Oil is extracted from the seeds and sold as an expensive (and unproven) dietary supplement – or you could just add them to dishes like stews for a nutritional boost.

A word of warning, however: all parts of the plant can cause a scratchy sensation in the back of the throat when eaten raw. Sam Thayer describes them as 'spicy'[9] and this quality doesn't seem to bother some people. Personally, while I'm a big fan of chilli heat and the sinus-blowing effects of mustard, I don't think I'll ever grow to like the effect

Evening primrose flowers

of evening primrose 'spice', which feels far more like a warning of toxicity. Cooking eliminates the effect in the aerial parts but not entirely in the roots.

There are many other members of the genus *Oenothera*, both biennial and perennial. A lot of them have been recorded as edible and would likely be worth trying.

Opium poppy
Papaver somniferum

PAPAVERACEAE

I once walked into my allotment to find a Bangladeshi friend looking quizzically at my poppies and asking, "Are these poppies… like in Afghanistan?" *Papaver somniferum* is indeed the infamous opium poppy, but the subspecies *hortense* found in British gardens produces negligible levels of opiates, so hopefully your forest garden won't be raided by the police. Besides their pharmaceutical uses, poppies have been grown for centuries for their abundant, oil-rich seeds. These are borne in large, round seed heads which develop a pleasing pepper-shaker arrangement as they ripen and dry. Pores around the plate at the top open up and seeds cascade out when the seed head is turned upside down. This makes them easy to harvest by picking the seed heads into a tub and allowing them to roll around.

Hoverflies on the opium poppy cultivar 'Danish Flag'

Seed head developing in a 'pom-pom' variety

Ripening poppy seed heads

Cultivars which do not open up are also available; these minimise seed loss but take more work to crush the capsules and separate the seed from the resulting chaff.

The seeds are traditionally used in bread-making, in korma curries and as a paste in some pastries, and they can also be added to soups and stews. When fresh they have a soft texture and a delicious nutty taste and my favourite way to eat them is straight from the plant, into the hand and into the mouth. The oil that gives them their moist texture and nutty taste is easily lost and older seeds become dry, somewhat gritty and less immediately flavoursome. They can also easily go rancid. I have found that the best way to preserve them is to keep them in jars in the fridge, and it would probably be worth freezing some if you had lots.

Poppies are, of course, also strikingly attractive flowers, and hoverflies in particular are very fond of them. Most of the ornamental strains are just as good at seed production, so you don't have to choose between beauty and food.

Rhubarbs
Rheum species

POLYGONACEAE

Rhubarb is the one perennial vegetable that is in everyone's garden, if only because, once it is planted, little short of a direct nuclear strike will shift it again. Gardens abandoned for decades can still have healthy patches of rhubarb. The cultivated rhubarb (*R*. x *hybridum*) is, as the name suggests, a hybrid species with an increased chromosome count. Its ancestors appear to be the wild rhubarbs *R. rhabarbarum*, from Siberia and northern China, and *R. rhaponticum*, from southwest Europe.

Due to its tart flavour, rhubarb didn't really take off as a food until the advent of cheap sugar, which enabled use of the leaf-stalks as a sort of honorary fruit. Indeed, my earliest memories of it are of eating sticks of it raw, dipped in a bowl of sugar. It is good at carrying the flavours of other ingredients,

Blanched rhubarb stems contrasting with unblanched ones

Rhubarb flowers

and as a jam it is best mixed with something like elderflower (see elder, Chapter 7) or ginger. It also makes a great chutney (see box overleaf). Combining it with pastry or similar also counteracts the sourness, as in rhubarb crumble or pie. Rhubarb pie is heavenly the day after making, once the juice has had time to soak into the pastry! It is also one of the better and more reliable country wines.

There is more, however, to rhubarb than using it as a pseudo-fruit. I frequently use a stalk as an ingredient in a stir-fry, added towards the end as it doesn't need much cooking. Stems can be made somewhat milder by blanching. The flower stems can be candied like angelica (Chapter 5), and the immature flowers can be used as a vegetable (with as much of the stem, which is more sour, removed as possible).

Finally, what of the leaves? Thomas Jefferson planted *R. undulatum* (as synonym of *R. rhabarbarum*) at Monticello, and described it as "Esculent rhubarb, the leaves excellent as Spinach".[10] In contrast, I – like most people – grew up 'knowing' that the leaves of rhubarb were deadly poisonous due to their super-high oxalic acid content. In reality, the leaves don't contain much more oxalic acid than the stems, and less than a range of herbs and vegetables, like carrot, parsley and spinach,[11] that we eat on a regular basis (although bear in mind that these are average values and individual plants vary). The belief seems to stem from a number of deaths that occurred after the public were encouraged to use rhubarb leaves as a substitute for spinach during World War I. These deaths may have been from a toxin other than oxalic acid in the leaves. Anthraquinone from the roots of rhubarb was used first in Chinese and then in European medicine as a purgative and laxative and has been described as 'The All-Bran of the Age of Reason'.[12] The deaths may conceivably have been caused by the consumption of extremely large quantities of leaves, perhaps on a regular basis. Oxalic acid is indeed poisonous in large quantities, but the lethal dose for a 65kg (10st) human is around 5kg (11lb) of leaves. The final alternative is that the deaths may have been caused by something else and simply misattributed to rhubarb. Since no-one eats the leaves anymore we don't have the data to find out. My own cautious experiments suggest that, in any case, the question isn't really worth answering as they don't taste very good! You can, however, safely ignore subsidiary myths about the awesome toxicity of rhubarb leaves, such as the belief that they shouldn't go on the compost heap or that green stems shouldn't be eaten.

A number of other rhubarbs are available. Himalayan rhubarb (*R. australe*) is said to taste of apple but in my experience isn't hardy enough for an

Turkish rhubarb, taken before the flower heads disappeared out of frame

RHUBARB CHUTNEY – THE REBOOT

I don't like traditional rhubarb chutney, made with curry powder, raisins and loads of sugar, very much. This reimagining of it plays to the strengths of rhubarb more, with less sugar and a cleaner taste, plus other ingredients from the forest garden. The vinegar is added right at the end to avoid boiling most of it off and give more control over the acidity.

Ingredients

2kg (4.4lb) rhubarb stems
8 sweet cicely shoots
2 lovage shoots
2 alexanders shoots
1kg (2.2lb) sugar
30g (1oz) fresh root ginger
approx. 150ml (5 fl.oz) pickling vinegar

Method

1. Cut the rhubarb, sweet cicely and lovage into 1cm (0.4in) lengths, cover with the sugar and leave for a day or two for the sugar to draw all the juice out of the vegetables.

2. Put in a large pan. Chop the ginger finely and add. Cook until it has become thick and 'jammy' (takes 30-40 minutes).

3. Add the pickling vinegar, a bit at a time, until the balance between sweet and sour tastes right to you (it's hot, obviously, so don't scald your tongue).

4. Pour into heated, sterilised jars and seal immediately.

Aberdeen winter. Sikkim rhubarb (*R. nobile*) would be worth growing, if only for its curious habit of growing its own greenhouse. Translucent bracts on the flower stem interlock to create a shelter that raises the temperature inside, protecting the flowers from extreme cold and UV-B radiation.

A more practical option, however, would be *da huang* or Turkish rhubarb (*R. palmatum*). This comes from China but was first encountered by westerners in Turkey – this is the case with quite a lot of plants with 'Turkish' in their name. It is a spectacular plant, with enormous, deeply-cut leaves and a flower spike that grows to two metres and more and bears clouds of candy-pink flowers. It's also a good culinary plant. Its flavour is more complex than that of hybrid rhubarb, lending itself more to savoury uses. One advantage of it is that it is very hardy and grows on for a month or two after garden rhubarb has stopped producing.

Rhubarb being harvested at the Orkney Wine Company

Herb patience leaves outpacing the wild garlic in early spring

Herb patience flower shoots

Docks
Rumex species

POLYGONACEAE

Docks are closely related to the rhubarbs, sharing with them a perennial growth habit, thick, fleshy root and a sour taste from oxalic acid. You may spend more time taking them out of your home garden than putting them in, especially in the early years, but a few are worth cultivating deliberately. Some of these are 'sorrels', dealt with overleaf. Of the others, herb patience or patience dock (*R. patientia*) is definitely the best. It has very mild leaves which make a good pot herb, and a great-tasting flower shoot. My favourite way to cook it is to batter and deep fry it; it goes soft, almost gooey, inside the crispy batter and the slight sourness gives it a little bite.

The native curly dock (*R. crispus*) is not bad either, but such a common weed that it might not be necessary to grow it in your garden (or popular with your neighbours). It looks similar to the far more bitter round-leaved dock (*R. obtusifolius*) but easy to distinguish once you get your eye into its narrower, rather distorted leaves. Monk's rhubarb (*R. alpinus*) is a passable pot herb, but its aggressively spreading rhizome means that it is best literally confined to a pot.

Mature seed head of herb patience amongst summer growth

Sorrels
Various species

VARIOUS FAMILIES

Sorrel is a perennial vegetable that takes me back to my childhood. Roaming the hills above our house, thirsty from having forgotten to take water, I would seek out the apple-green leaves of *Rumex acetosa*. They did nothing to really stop me getting dehydrated I'm sure, but the rush of saliva brought on by the acid taste always made me feel better. We called them soories from the sour taste, but the more widespread name is common sorrel.

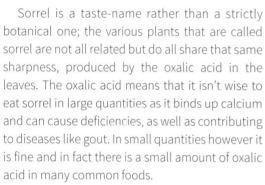

Wood sorrel

Buckler-leaved sorrel

Sorrel is a taste-name rather than a strictly botanical one; the various plants that are called sorrel are not all related but do all share that same sharpness, produced by the oxalic acid in the leaves. The oxalic acid means that it isn't wise to eat sorrel in large quantities as it binds up calcium and can cause deficiencies, as well as contributing to diseases like gout. In small quantities however it is fine and in fact there is a small amount of oxalic acid in many common foods.

Several of the sorrels are in the *Rumex* genus. Sheep's sorrel, *R. acetosella*, is tasty but too small and invasive to be worth cultivating deliberately. Buckler-leaved sorrel, *R. scutatus*, is also small-leaved, but very productive, and the leaves with their cool shape look great in a salad. My old friend common sorrel is the one that gardeners have paid most attention to and there are many cultivated forms, bred for larger leaves, usually sold as French or Polish sorrel. There is also a non-seeding form called 'Profusion' which has the advantage that it puts all its energy into producing leaves rather than running extravagantly to seed in the summer. You might also find a 'red-veined sorrel' in seed catalogues. This is the marketers' spin on bloody dock (*R. sanguineus*), a pot herb that looks very pretty but is only marginally edible.

Other sorrel-tasting plants are in the genus of *Oxalis*. This is a huge genus with around 800 species, variously known as sorrels (which they aren't), shamrocks (which they aren't) and grasses (which they aren't). Confused? You will be with common names. Species include the Andean crop oca (*O. tuberosa*), which is grown for its tubers but worth pinching a few leaves from for a salad; the native wood sorrel (*O. acetosella*), which grows in heavy shade but isn't very productive; and iron cross plant (*O. deppei*), so called for the resemblance of its leaves to the German Iron Cross medal.

You can also find the sorrel flavour in mountain sorrel (*Oxyria digyna*), a plant of damp, rocky mountain soils that struggles with the competition in the more fertile conditions of a home garden. There is even a sorrel tree, *Oxydendrum arboreum*, which is a member of the heath family, but a whole tree of sorrel might be too much of a good thing.

I mostly use sorrel chopped into a salad, but its most traditional use is in sorrel soup, particularly in Eastern Europe, where it is sometimes known as green borscht. Here's my favourite sorrel soup recipe.

SORREL SOUP

Ingredients

oil for frying
1 onion, 1 clove garlic, chopped
1 potato, cut up small
600ml stock (1 pint)
handful sorrel leaves, chopped
optional: 1 egg, beaten

Method

Fry the onion in the oil until soft, then add the garlic and the potato and fry for a couple minutes more before adding the stock. The stock can be whatever kind you like; chicken is traditional but I use vegetable. Bring to the boil and add the sorrel leaves, which will immediately lose their bright green colour and go much darker. Cook for around 10 minutes until the potatoes are soft, then blend and serve hot or cold.

The size of the handful of sorrel leaves depends on how sour you want it to be. One hundred grams (4oz) or more will give it a good bite; 50g (1.7oz) will give a hint of sorrel that slowly grows on the back of your tongue as you work your way through a bowl, an effect that I quite enjoy.

For a twist, you can add a beaten egg at the end, cooking for a few more minutes, which gives the soup a lot of body. Serve with a hard-boiled egg or a dollop of sour cream.

Potatoes
Solanum tuberosum

SOLANACEAE

I'm sure I don't need to tell anyone how to grow or cook tatties, but I would like to point out that the humble spud is in fact a perennial vegetable, and one well suited to growing in a home garden. Potatoes like a rich, well-watered soil, and their large overwintering parts (the potatoes themselves) and toxic leaves mean that they have no trouble getting started amongst other plants. I grow almost all of mine in small patches, with plenty of added compost, scattered through the sunny and woodland edge beds.

We typically grow potatoes like annuals, buying in new 'seed potatoes' every year, because of the risk of disease building up in saved tubers. The disease risk is lower in cool areas, and Aberdeenshire has a long history of growing seed tubers to be grown on in other areas. This means that with a combination of growing blight-resistant cultivars such as the Sarpo family and weeding out any line that shows signs of disease, you can grow potatoes more like a traditional perennial, saving your own tubers or even leaving some in the ground to sprout the next year. They still benefit from 'earthing up':

Amusingly-shaped potatoes are one of the joys of growing your own. I thought this one looked like Comet 67P/ Churyumov–Gerasimenko.

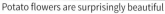
Potato flowers are surprisingly beautiful

raking soil or compost over the base of the plants to prevent the tubers from being exposed to light and turning green (and toxic).

You can also propagate your own potatoes by growing them from seed. True seeds (as opposed to 'seed' potatoes, which are tubers rather than seeds) are found in the tomato-like fruits that some potato varieties bear. They can be sown to produce tiny tubers, which bulk up over the next year or two to produce a more familiar-sized spud. Any plant propagated entirely by tubers inevitably builds up a disease load, and growing some from seed each year refreshes your stocks. You also have the excitement of breeding your own varieties, unique to your garden.

The potato is only one of a wide range of Andean tuber crops, most of which can be grown in similar ways. These include oca (*Oxalis tuberosa*), with little, potato-like tubers with a lemony taste; ulluku (*Ullucus tuberosus*), which have striking colours and the feel of unearthing buried treasure when you dig them up, but taste more prosaically of beetroot; mashua (*Tropaeolum tuberosum*), a climbing relative of nasturtium which tastes appalling to

begin with but is apparently much improved by slow cooking; maca (*Lepidium meyenii*), a relative of the wasabi-flavoured dittander; mauka (*Mirabilis expansa*), a trailing plant that grew at frightening speed in my allotment but produced barely any root); and yacon (*Smallanthus sonchifolius*), a sunflower relative which has both small vegetative tubers and large, sweet storage tubers.

All these crops are adapted to the cool conditions of the high Andes but share one major drawback when grown at higher latitudes. Used to the relatively unchanging conditions around the equator, they are in no hurry to produce tubers and only do so in the tail end of the year, prompted in many cases by shortening day lengths. This means that they are often caught unawares by frost when they have put little of their resources into tubers, and crops can be minimal in years with an early frost. Potato was once like this too, until a mutation was discovered that caused plants to start forming tubers far earlier in the season. It is reasonable to hope that this might be achieved with others of the Andean crops, giving us more starchy tubers for home gardens in the future.

Dew on nasturtium flowers

Nasturtium
Tropaeolum majus

TROPAEOLACEAE

It is well known that nasturtiums are edible, and their pretty, cress-flavoured flowers are often added to salads. What is less well known is that their leaves, cooked like spinach until the volatile oil that gives them their peppery bite is driven off, are also quite a delicacy, with a flavour completely different from the raw one. The tips of the shoots (about 10cm/4in) are nice in stir-fries and the green seeds can be pickled like capers.

Nasturtiums are trailing annual plants that scramble over supports or other plants. They put on most growth quite late in the season, which means that they can be combined with anything that dies down early to make best use of the ground. They are not frost-hardy and are usually cut down by the first frost. They self-seed in disturbed ground or they can be helped out by growing young plants in pots and planting them into a cleared patch.

Wet ground

A final factor that can lead to open habitats is wetness. Plants of open water are not included here as few people have the size of pond to grow them in significant quantities, but it is far easier to create (or leave) an area of wetland for plants that like permanently wet ground.

Sweet flag
Acorus calamus

ACORACEAE

Sweet flag is a plant of muddy riversides, with creeping rhizomes that throw up a crop of flattened leaf bundles ('flags') every spring. Its rhizome has been used around the world for a myriad of medical purposes, but the edible parts are the tender 'hearts' at the base of the flags in spring and the immature flower spikes. The whole plant is suffused with a flavour that reminded me slightly of marzipan but is really unique; even chewing a fragment of stem while working in the garden is pleasant. Sam Thayer describes the flavour of the flower spikes in stellar terms; mine, tantalisingly, haven't produced any yet.

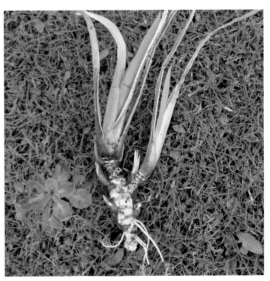

Acorus calamus rhizome with new buds

Water parsnip
Sium sauve

APIACEAE

Water parsnip is a relative of skirret (p.248) and is similar in its uses. It grows in standing water or muddy soils with lots of organic matter; if you don't naturally have these conditions it is probably easier just to grow skirret! In the past water parsnip was an important food of the Cree nation.

Chufa
Cyperus esculentus

CYPERACEAE

Chufa is in the same genus as papyrus; the plant used by the Ancient Egyptians to make paper. Its useable part is the little tubers that are borne in abundance on the roots. They have both the taste and texture of coconut flesh, and can be blended with water to make the delicious drink horchata. Under the marketing term 'tiger nuts' they have become something of a health fad.

Chufa is naturally a wetland plant, but will succeed in ordinary soil if well watered. The trickiest part can be getting tubers to sprout in the first place; I find that soaking them well and starting

Chufa tubers

them in a propagator with some bottom heat is the most reliable method.

C. esculentus comes in both wild and domesticated forms. Tubers bought from the shops will be the domesticated type, which has larger tubers but rarely seeds and isn't 100% happy in temperate gardens. In North America, there is a wild form known as nutsedge or yellow nutsedge, which grows as far north as British Columbia and Nova Scotia. It produces more seed and smaller tubers and is a serious agricultural weed in some parts. It would be nice if someone could cross the two to make a plant with the tuber size of the cultivated form and the vigour and hardiness of the wild one.

Water mint
Mentha aquatica

LAMIACEAE

Water mint is a wonderfully aromatic mint adapted to growing in streams and pond edges. Like other mints, it can be somewhat invasive. It is one of the parents of peppermint (*Mentha* x *piperata*).

Woundworts
Stachys species

LAMIACEAE

The tubers of Chinese artichoke (*S. affinis*) look so much like big, fat, white maggots that I once managed to convince a flatmate that that was what we were having for dinner that night. It may say something about how many weird foods I served up that she believed me – or maybe she was just humouring me. The 'artichoke' in the name refers to the flavour, which genuinely is reminiscent of that of globe artichoke. It has also become common recently to call the plant *crosne*, from the name of the commune where it was first grown in France.

The biggest drawback to Chinese artichoke is that the knobbly tubers are excellent at holding on to dirt. They are not large, so cleaning them can be quite a faff. One trick is to swirl them with sand in a

Chinese artichoke (left) and marsh woundwort (right) leaves. Marsh woundwort leaves are quite variable and can also look similar to crosne's.

Chinese artichoke flowers

Marsh woundwort tubers

Chinese artichoke tubers

bucket of water. My normal technique is to dump a batch in water, then clean them with a toothbrush, working from the largest to the smallest until I get fed up. The remainder then go to the compost.

Another drawback is that, according to most sources, Chinese artichokes are sterile and do not produce viable seed. Fortunately no-one has told my plants, which produce seeds (especially in hot years) that I have successfully grown on into new plants. It normally reproduces, however, by means of the tubers, some of which are inevitably left in the soil after harvest.

Chinese artichoke will flourish in any rich, well-watered soil, but its relative marsh woundwort (*S. palustris*) is a genuine wetland plant. It produces longer tubers but fewer of them. It is a native Scottish plant, so you could try collecting seed locally. In North America, according to the Eat the Weeds foraging website, hyssopleaf hedgenettle[13] (*S. hyssopifolia*) is similar.

Bog myrtle
Myrica gale

MYRICACEAE

Bog myrtle, also known as sweet gale, is a hardy sub-shrub growing in wet places around the northern hemisphere. Perhaps its greatest claim to fame,

Bog myrtle

Averons growing at 1,000m above sea level in Norway

in Scotland at least, is as a midge repellent. I'm not sure if the midgies know this, but there is often some available where they are their worst, and crushing some leaves to rub on your face or putting a sprig behind your ear at least takes your mind off them for a while and makes you smell nice.

Almost as importantly, it was used in the past as a flavouring for beer, similarly to hops, and a number of craft brewers are rediscovering it in this role. Its taste is sweet, aromatic, resiny and a little astringent. If you aren't making beer, dried sprigs can also be used as a food flavouring like bay leaves. If all these virtues weren't enough, it is one of the few plants outside the bean family able to fix its own nitrogen, helping it to flourish in boggy, nitrogen-poor soils.

Averons
Rubus chamaemorus

ROSACEAE

Averons, also known as cloudberry or bakeapple (from the French, baie qu'appelle) is a low-growing member of the raspberry genus which makes a living trailing across the surface of deep peat. This is a habitat you are unlikely to have in your garden, but it also grows in native pinewoods, so it should be worth trying in a well-watered spot in a forest garden.

Cloudberries are particularly popular in Scandinavia, where they are widely foraged for jam, liqueurs and desserts. A few improved forms for cultivation have also been developed there. Plants are normally diooecious, so a male form such as 'Apolto' or 'Apollen' is needed to pollinate female cultivars like 'Fjordgull' and 'Fjellgull'. The cultivar 'Nyby' is hermaphrodite, with both male and female flower parts.[14]

Fish leaf
Houttuynia cordata

SAURURACEAE

Houttuynia cordata is native to Southeast Asia and features in the cuisines of parts of India, Vietnam and southern China, but despite this southerly origin it grows well in cooler climates and has been widely adopted in British gardens. This means that it is most often available as ornamental cultivars. Multi-coloured variegated forms like 'Chameleon' are striking but grow more slowly than the species. 'Flore pleno' has double flowers and green leaves and grows as strongly as the species.

Both leaves and rhizomes (which the plant uses to spread through wet soils) are eaten. The leaves make a strongly-flavoured herb, often described as tasting like orange peel, but in Asia more often

Leaves and flowers of *Houttuynia cordata* 'Chameleon'

likened to fish, with names like fish mint or fish leaf. They can be cut finely and added to salads or cooked dishes. The rhizome can be washed, chopped and used either raw or cooked, in everything from salads to fried rice.

Reedmaces
Typha species

TYPHACEAE

Reedmaces are also known as cattails and bulrushes, although this last name is perhaps better not used so as to avoid confusion with a group of large sedges that go by the same name. Their instantly recognisable seed heads, like a sausage skewered lengthways on the end of the flower stem, make them a visual shorthand for pond vegetation in pictures of all kinds.

Great reedmace (*T. latifolia*) is found widely through North America, Eurasia (including Scotland) and even Africa. It is a very vigorous species and pond enthusiasts usually grow it in baskets to contain the rhizome and stop it taking over the entire pond. It is usually found in open water but will also grow in pond margins or any rich, permanently wet soil. Lesser reedmace (*T. angustifolia*) is smaller and less widespread. The two species sometimes cross to form a sterile hybrid.

Reedmaces have a host of uses (leading to the nickname 'the supermarket of the swamps'), although not all of them are worth the effort required. According to Sam Thayer, the best part is the young, growing rhizome, which is sweet, soft and mild flavoured and can be used cooked or raw.[15] When the rhizome matures it becomes tough and fibrous; starch can be laboriously extracted from it, but with the quantities that can be produced in any normal sized garden this is really not worth it. In spring to early summer, the 'heart' of the shoots (the tender bases of the inner leaves) provide another tender vegetable, sometimes called 'Cossack asparagus'. When the plants flower, the tip of the flower head (the male part) can be steamed and eaten like corn on the cob. If this part is left it will produce pollen, which can be collected and used as a (non-sticky) flour like pine pollen. If you get enough to store, it needs to be dried before storing.

A fruiting shiitake log

12

Fungi

Alongside the plants, a forest garden also provides the conditions that many fungi like best: cool, moist and with plenty to decompose. Useful fungi broadly fall into two categories: the mycorrhizal species which grow in symbiosis with plant roots, and the saprophytes, which make a living by breaking down dead plant material. Many of the best mushrooms, such as ceps or chanterelles, are mycorrhizal, and unfortunately there is still no reliable way of cultivating these – or perhaps fortunately: it is nice still to have an excuse to venture out into the woods and enjoy foraging these hidden treasures.

The saprophytes are far easier to tame; you just have to introduce the fungus to a food source that it likes and provide suitable conditions. Here I'll stick to those that can be grown on wood or compost, in situ in the forest garden – although many more can be grown if you import substrates such as manure for them. There are two main possibilities: growing on logs and on beds of woodchip. For either, you'll need to make or buy 'spawn' of the appropriate species. This is simply the fungus grown on a suitable substrate such as grain, bran, sawdust or even wooden dowels. You can often see the hyphae, the 'body' of the fungus, as a mass of white strands holding the spawn together.

Growing on logs

For log-growing, sourcing good logs to grow the fungus on is probably the hardest part of the whole process. Most of those listed below will grow on a wide range of broadleaf species; only a few (Indian oyster mushroom and bear's head) will grow on conifer logs. Dense woods like beech, oak, elm and lime last longer: I have had six years of shiitake fruiting out of a good beech log. Lighter woods like birch, aspen and willow are also good but won't last so long. Sycamore is a good species in many respects, but has an unfortunate tendency for its bark to fall off in sheets, after which it becomes much harder to prevent the log from drying out and dying.

Time of year is important too. As with firewood, the ideal time to cut mushroom logs is in the winter months, before the sap run starts. When the sap begins to run, the bark becomes softer and looser to accommodate the growing wood. This makes it more likely to fall off, leaving the log to dry out and the fungus to die. Once this initial burst of growth is over, logs cut in the summer will generally be okay too.

Size and shape also matter. Species that often grow on stumps, such as pearl oyster, hen of the woods and chicken of the woods, can be grown on large logs, partly sunk into the ground, or indeed inoculated straight into a fresh stump. With the others, the main thing is that the log should be at least 10cm (4in) in diameter and half a metre (1.6ft) long, so that it doesn't dry out too fast. Above these dimensions, the main limitation is that you have to be able to handle it and, with shiitake, fit it into some sort of water bath (more on that later). Straight logs are easier to store and handle, but there's nothing stopping you using crooked ones: the fungus won't mind. Some fungi can usually only digest the lighter-looking sapwood in a log, and not the darker heartwood that forms in the middle. This means that using particularly large logs with a lot of heartwood is a bit of a waste. It also means that very slow-growing trees don't have much on even a small log to interest the fungus, which can be a problem with trees like oak growing in less than ideal conditions.

The other major consideration is that a log mustn't provide too many entry points for 'weed' fungi. Cultivated fungi can suffer from competition with opportunistic species that turn up and try to take over the log, just like plants in a bed. Damaged bark, cut branches and pockets of rot all provide entry points. A few are not a problem, but lots will require extra materials to inoculate, making the process more expensive.

It is often recommended to get your mushroom logs from a tree surgeon, but in my experience tree surgeons are rarely interested in producing logs of the right length, carefully handled to avoid damaging the bark. If you have your own wood then cutting mushroom logs can be a perfect part of thinning to produce a broadleaved timber crop. If your forest garden is large then you can even plant suitable species and harvest them by coppicing (certain species can be cut close to the ground, sending up new shoots that become the next tree) for an abundance of logs. The rest of us have to come to an arrangement with a landowner or be more opportunistic. A wind-thrown beech can be a bonanza as the owner is sometimes happy to have it removed from where it has fallen.

The logs then have to be inoculated, a process of drilling holes, filling them with some form of mushroom spawn and then sealing both the drill

Fresh logs and mushroom spawn

holes and any damaged bark with molten wax. Anyone selling mushroom spawn will generally give detailed instructions on this and happily sell you the specialist bits of kit that make the job easier. What is less well covered is how to store the log afterwards. I suspect that the majority of bought logs die before fruiting due to unsuitable storage. After losing some myself I now have a system that works. I have a small wooden bay built in a shady corner of the garden, which allows me to stand the logs vertically, and a piece of shade netting over the top. In dry periods over the summer I wet them once a week or so. In conditions like this, most fungi will colonise the wood within about a year, and may be visible as triangular, brown or white, mushroom-scented patches at the end of the log.

Once a log is colonised, it is ready to fruit. Most species fruit when they feel like it, but shiitake have the advantage that an individual log can be induced to fruit whenever you choose. This allows fruiting to be staggered instead of coming in one great glut and makes shiitake, in my opinion, the most useful species of all. The fruiting process is described overleaf.

A colonised log, showing patches of mycelium along the grain of the wood from the inoculation point

Growing in beds

Another option is to grow fungi in beds of a substrate such as woodchip. A mushroom bed needs a shady spot, so underneath a tree is perfect. Species that can be grown in this way include wine caps, morels and the adaptable oyster mushrooms. The general method, with tweaks for individual species, is to layer substrate onto the ground with spawn of the desired fungus.

The ground below should be reasonably free of weeds, to stop them from coming up through the bed. A layer of cardboard can be laid first as an added weed barrier. The next step is to add a layer of substrate, around 5cm (2in) thick. The substrate should not be too heavily colonised by other fungi, or the one you are cultivating will have a job getting established. If you are using woodchip, there are several ways of achieving this. The first is to use fresh chips, which haven't had time to colonise. The drawback here is that fresh chips will heat up as they break down, so you need to spread them thinly to avoid cooking your spawn. The gold standard is to sterilise the substrate. In a garden situation this can be achieved by heating the chips together with water in a metal drum – taking plenty of care with the hot water and metal! Put the woodchip in a wire basket inside so that it can be removed easily. Mycologist Paul Stamets recommends two hours of cooking at between 65 and 80°C (149-185°F).

A less certain but lower-input and decidedly safer method is to simply soak the chips for a week. Competing fungi will die during the soaking and the anaerobic bacteria that take their place will die after the chips are drained. Finally, you can just take your chances with untreated woodchip. Fungi grown in this way are generally strong competitors and might win out anyway. However, your chances of success will definitely be lower.

Once the first layer is down, the mushroom spawn is scattered over the top of it, then another layer of substrate is added. This is called the sandwich method for obvious reasons. Alternatively, the fungus may be supplied on dowels. Ann Miller, of Ann Miller's Speciality Mushrooms,[1] sells wine cap this way and recommends pushing the dowels lightly into the soil, leaving most of the length sticking up into the layer of mulch.

Care for a mushroom bed mostly consists of watering it during dry periods and making sure that it doesn't become taken over by plants. Once it is colonised, you can top it up with further layers of substrate to keep production going for longer. One method that I mean to try is inoculating my deep-wood paths. Fungi do not generally like compaction so the centre of the path is unlikely to be good territory for them, but it would be interesting to find out to what degree a bed established at the edge of a path was able to colonise it.

Even if you do nothing to bring fungi into a forest garden, the hospitable conditions that you create may lead to some turning up of their own accord, and it is worth keeping an eye out for a chance to forage mushrooms in your own garden. I have found blewits and the disturbingly brain-like *Gryomitra esculenta* in this way.

Species

Shiitake
Lentinula edodes

ORDER: AGARICALES

Shiitake have been cultivated in China since at least the Song dynasty, 800 years ago. They succeed on a wide range of hardwood logs and grow strongly to outcompete any 'weed' fungi that may try to get into the log. Colonised logs can be induced to fruit at any time of year by 'shocking' them. This involves beating them (hard enough to reverberate through the log but not so hard that you damage it) with a lump hammer, then submerging them in water for a day or two. In summer and autumn the logs can be left outside to fruit, although they do need to be protected from slugs and snails, which

love the mushrooms as much as we do. I put mine in my shed. In winter and spring they need to be taken indoors where it is warm. Mushrooms appear within a week and are ready to pick within two.

Shiitake are delicious. They are very versatile, but one of my favourite ways to use fresh ones is to cut them thin and fry them well, upon which they take on the texture of crispy bacon. Any that you don't use fresh will dry if left on a sunny windowsill and keep well. They can be rehydrated later for use in soups and stews.

Oyster mushrooms
Pleurotus species

ORDER: AGARICALES

Oyster mushrooms have a mild flavour and good texture and need little cooking. They are great in stir-fries and Italian dishes. The most commonly used species is pearl oyster (*P. ostreatus*), which grows on a range of hardwood species. It is currently flourishing in Aberdeen, taking advantage of the glut of dead elm trees caused by Dutch elm disease. They fruit mostly in the late autumn and winter, helping to extend the mushroom season. They have broad tastes and have been grown on a wide range of substrates, including sawdust, straw, coffee grounds and even toilet rolls or old books. Indian oyster mushroom (*P. pulmonarius*) is useful as it also grows on conifer logs such as pine. It has thicker flesh and fruits earlier in the year than the pearl oyster.

Wine cap
Stropharia rugosoannulata

ORDER: AGARICALES

Wine cap is a large mushroom, sometimes known as king stropharia, garden giant or even Godzilla mushroom in honour of its size. The caps are a deep, rich burgundy colour. It is best grown on thick beds of mulch, in which you can include hardwood sawdust, straw, woodchip, torn-up cardboard and shredded woody prunings from the rest of the

Drying shiitake shrink considerably!

garden. Newly-made beds must be soaked well and kept wet for a couple of days. Wine caps are fast colonisers and should appear the same year if beds are made in the spring. Despite the size of the mature mushroom, they are best picked when still only fist-sized.

Velvet shank
Flammulina velutipes

ORDER: AGARICALES

Velvet shank is also known as winter mushroom for its ability to survive freezing solid. This may be the only mushroom to have gone into space: in 1993 samples were flown on the Space Shuttle Columbia in order to determine how they would handle low gravity. The white, thread-like 'enoki' that you see on sale in supermarkets are actually the same fungus, but grown in a cold, dark environment with high CO_2 levels that mimic conditions underground, making the fruiting bodies stretch madly looking for the surface. In the forest garden it is grown on hardwood logs.

Jelly ears
Auricularia auricula-judeae

ORDER: AURICULARIALES

Jelly ears are unusual in a number of ways. They are the only fungus I know of that is particularly fond of elder, often growing on dead snags within an otherwise healthy tree. You can't buy spawn, but bring a fruiting stick into your garden and soon you'll have them popping out of every bit of dead wood around. Their fruiting bodies are long-lived and seem impervious to both drying out and freezing solid, returning to a soft, jellyish texture as soon as milder conditions come back. They don't taste of much, but can be used as a textural addition to stir-fries or miso soup, or used to carry flavour by rehydrating fried caps with something tastier. The forager and chef Fergus Drennan came up with the idea of soaking them in elderberry liqueur and then partially drying again to make an unusual variety of Turkish delight.[2]

The Latin name hints at a darker past for this fungus. They used to be known as Jew's ears or

Judas ears, due to a piece of anti-Semitic mythology associated with the elders that they grow on. These days 'jelly ear' or 'wood ear' is definitely to be preferred.

Morels
Morchella angusticeps

ORDER: PEZIZALES

Morels are poorly understood and difficult-to-grow mushrooms, but worth persisting with for their rich, meaty taste and striking looks. There are a number of edible species, some of which are thought to be mycorrhizal and some of which are saprophytic. Any cultivated varieties you can source will obviously be saprophytic. Ann Miller offers a strain of *M. angusticeps* originally found in Speyside. Many species have an association with fire, and grow well in the alkaline conditions created by wood ash. Ann recommends planting the spawn in an old fire site. Indoor growers use a substrate mixed from 75% sterilised organic material including potting compost, woodchip and ground bark, 25% sand and a small amount of lime to make the mixture alkaline. The original patent for growing morels commercially suggests that successfully colonised beds can be shocked to induce fruiting by giving them a thorough soaking.[3]

Birds of the woods
Grifola frondosa and *Laetiporus sulphureus*

ORDER: POLYPORALES

Maitake, or hen of the woods (*G. frondosa*), looks like an old grey hen fluffing out her feathers and is renowned for its rich, fungal flavour. Staying with the poultry theme, chicken of the woods (*L. sulphureus*) is named for its chicken-like texture rather than its appearance, which is bright yellow. They are both polypore mushrooms which like growing on stumps, so you need a large log and a shady spot where you can sink it into the ground.

Lion's mane and bear's head
Hericium erinaceus and *abietis*

ORDER: RUSSULALES

Lion's mane and bear's head are two fungi in the genus *Hericium*, sharing an appearance like a mass of delicate, whitish stalactites and a fish-like taste. Lion's mane grows on hardwood logs, especially beech, while bear's head (also known as coral fungus) prefers softwoods such as firs (*Abies*), Douglas fir (*Pseudotsuga menziesii*) and pines.

Endnotes

Chapter 1

1   Nie *et al.* (2019)

Chapter 6

1   Hu (2012) Kindle Locations 880-881
2   Hu (2012) Kindle Location 887
3   foragerplants.blogspot.com
4   Hu (2012) Kindle Locations 3589-3592

Chapter 7

1   Cox and Beaton (2012) p.103
2   Cox and Beaton (2012) p.102
3   Cox and Beaton (2012) p.108
4   Cox and Beaton (2012) p.109
5   foragerplants.blogspot.com/2018/06/
    bird-cherry-prunus-padus
6   Thayer (2006)
7   foragerplants.blogspot.com/2018/06/
    bird-cherry-prunus-padus
8   Turner (1979) p.147
9   Turner (1977) and Turner (1979)
10 Wikipedia
11 Plants For A Future
12 Cox and Beaton (2012) p.110
13 Brussell (2004)
14 kanon1001.web.fc2.com/foto_sinrin/K_ukogi/
    takanotume/takanotume
15 Irving (2009) p.202
16 Irving (2009) p.203
17 Glasse (1747), Bradley (1750) and Eaton (1822)
18 Thayer (2010) p.405
19 Martins *et al.* (2014)
20 Mithen *et al.* (2001)
21 Cox and Beaton (2012) p.117
22 pfaf.org/user/Plant.aspx?LatinName=Caragana+
    arborescens
23 www.facebook.com/groups/
    PlantBreedingForPermaculture/
    permalink/1259182987578665
24 www.facebook.com/groups/
    PlantBreedingForPermaculture/
    permalink/1259182987578665
25 Turner (1997) p.56
26 Turner (1997)
27 McGee (2007)
28 pfaf.org/user/Plant.aspx?LatinName=Zanthoxylum+
    simulans
29 jurassicplants.co.uk/collections/all-plants/products/
    zanthoxylum-bungeanum-large-fruited-sechuan-pepper-
    tree

Chapter 8

1   Barstow (2014) p.135
2   Barstow (2014) p.137
3   Coffey (1993)
4   permies.com/t/14692/
    Aralia-Racemosa-American-Spikenard
5   Brussell (2004)
6   Thayer (2010) p.278
7   Thayer (2006) p.321
8   Barstow (2014) pp.102-109
9   chestnutherbs.com/sweet-shrub
10 Howell (2019)
11 Turner (1997) p.74
12 Moerman (2010) p.243
13 Thayer (2017) p.266
14 Thayer (2017) p.274
15 Thayer (2017) p.273
16 lewisbamboo.com/edible-bamboo-species
17 www.guaduabamboo.com/blog/edible-bamboo-species
18 Crawford (2010) p.176
19 Kang *et al.* (2012)

Chapter 9

1   www.youtube.com/watch?v=mI1fuqzNg2s
2   toads.wordpress.com
3   Thayer (2010) p.350
4   Barstow (2014) p.146
5   Thayer (2010) p.97
6   Turner (1997) p.75
7   Barstow (2014) p.201
8   Barstow (2014) pp.229-231
9   Barstow (2014) p.230
10 Martynoga (2012) p.46
11 Svanberg (2012)
12 Turner (1995) p.69
13 Hayward (1891) 'Popular names of American plants.'
    Journal of American Folklore 4 pp.147-50, quoted in Coffey
    (1993) p.310
14 Bergen, Fanny D (1894) *Popular American Plant Names*
    quoted in Coffey (1993) p.310
15 Plants For A Future database
16 Gibbons (1962)
17 Mears and Hillman (2007) p.117
18 grendz.com/pin/415/ and nakazora.wordpress.
    com/2012/04/21/this-month-we-are-mostly-eating
19 Thayer (2006) p.82
20 Medsger, OP (1939) *Edible Wild Plants*, quoted in Coffey
    (1993) p.13
21 Irving (2009) p.73
22 Mac Coitir (2003, 2015) p.62
23 Irving (2009) p.225
24 See, for example, nettlebeer.com
25 Wilde (2018)

26   Thayer (2010) pp.177-181
27   Barstow (2014) pp.188-190
28   nakazora.wordpress.com/2012/08/27/summer-in-the-shade, www.botanic.jp/plants-aa/aomizu and shizuokagourmet.com/2014/01/27/sansaiedible-wild-japanese-mountain-vegetables-2014-edition
29   Plants For A Future database

Chapter 10

1   Barney (2016)
2   Barstow (2014) p.211
3   Hu (2012) Kindle Location 3430
4   Hu (2012) Kindle Locations 4109-4120
5   Ladio & Lozada (2004)
6   Phillips (1983)
7   Martynoga (2012) p.57
8   Irving (2009) p.125
9   Turner (1995) p.58 and Turner (1997) p.79
10   Thayer (2010) p.371
11   Turner (1997) p.79
12   Hu (2005) *The Food Plants of China*, quoted in Thayer (2010) p.373
13   Hu (2012) Kindle Location 4403
14   Turner (1997) p.80
15   Jacke, Dave and Toensmeier, Eric (2015) *Edible Forest Gardens*, quoted in Plants For A Future database
16   bburikitchen.com/cham-namul-pimpinella-brachycarpa
17   Flora of China www.efloras.org/florataxon.aspx?flora_id=2&taxon_id=200015772
18   For a comparison, see http://digitalchosun.dizzo.com/site/data/img_dir/2019/05/30/2019053080140_1.jpg The labels read 참나물(chamnamul) and 파드득나물(Pa-deu-deug-na-mul, Cryptotaenia japonica)
19   Thayer (2006) pp.281-289
20   Barstow (2014) pp.250-255
21   Moerman (2010) p.50
22   Smith (2013)
23   www2.ville.montreal.qc.ca/jardin/en/info_verte/feuillet_fleurs_comes/tableau, accessed 7/5/2012
24   Turner (1997) pp.63-64
25   Hu (2012) Kindle Location 1914
26   daylilybreeder.blogspot.com/p/the-daylily-as-permaculture-subject.html accessed 8/6/2020
27   honest-food.net/dining-on-daylilies accessed 8/6/2020
28   daylilybreeder.blogspot.com/p/the-daylily-as-permaculture-subject.html accessed 8/6/2020
29   Moerman (2010) p.50
30   en.wikipedia.org/wiki/Doellingeria_scabra#Korea accessed 15/4/2020
31   en.wikipedia.org/wiki/Chwinamul 15/4/20
32   Kunkel (1984)
33   www.ediblewildfood.com/new-england-aster
34   Coffey (1993) p.247
35   Moerman (2010) p.58
36   nakazora.wordpress.com/2012/04/21/this-month-we-are-mostly-eating
37   Thayer (2017) pp.346-356
38   Barstow, www.edimentals.com/blog/?p=22018 accessed 8/3/2021
39   Barstow (2014) p.239
40   Thayer (2017) p.426
41   Hu (2012) Kindle Locations 3111-3112
42   bburikitchen.com/shiraegi-dried-radish-greens

43   Hu (2012) Kindle Locations 4146-4195
44   www.maangchi.com/recipe/doraji-muchim
45   Barstow (2019)
46   Hashimoto (2007)
47   Thayer (2006) p.242
48   Yıldırım *et al.* (2001)
49   Turner (1997) p.74
50   www.eatweeds.co.uk/wild-celery-common-mallow-harira
51   Ghirardini *et al.* (2007)
52   www.fondazioneslowfood.com/en/slow-food-presidia/ulleung-island-sanchae
53   e.g. www.facebook.com/guiaderecolecciontdf/photos/frambuesa-salvaje-rubus-geoides-tambi%C3%A9n-llamada-mi%C3%B1e-mi%C3%B1e-frutilla-silvestre-fru/2327266440869783 and waca.cl/tureco/2017/12/06/especies-comestibles-en-torres-del-paine
54   Cordero *et al.* (2017)
55   www.chileflora.com/Florachilena/FloraSpanish/HighResPages/SH0344
56   www.brc.ac.uk/plantatlas/plant/rubus-arcticus
57   Ryynänen (1973)
58   www.korewildfruitnursery.co.uk/RubusArcticus accessed 2/10/2020
59   Linnaeus (1737)
60   Wikipedia accessed 2/10/2020
61   www.uhi.ac.uk/en/t4-media/one-web/university/research/res-themes/agronomy-institute/pdf/Agronomy-Institute-Annual-Report-2006-2007.pdf accessed 8/3/2021

Chapter 11

1   Bryant (1783)
2   Turner (1997) pp.81-87
3   klamathsiskiyouseeds.com/wp-content/uploads/2019/02/KSNS_Seed_Germination-Chart_2019-Sheet1.pdf accessed 3/11/2020
4   Barstow (2014) pp.125-126
5   nakazora.wordpress.com/2012/04/21/this-month-we-are-mostly-eating accessed 21/10/2020
6   www.edimentals.com/blog/?page_id=1940
7   Barstow (2014) pp.165-173
8   Thayer (2006) p.278
9   Betts, Edwin M, ed (1944) *Thomas Jefferson's Garden Book*, 1766-1824: With Relevant Extracts from His Other Writings, quoted on www.monticello.org/site/house-and-gardens/rhubarb, archived 20/06/2011
10   en.wikipedia.org/wiki/Oxalic_acid and www.gardenmyths.com/oxalic-acid-rhubarb-leaves-harm-you, accessed 17/10/2020
11   www.gardenmyths.com/oxalic-acid-rhubarb-leaves-harm-you, accessed 17/10/2020
12   Barceloux (2008) p.84
13   www.eattheweeds.com/florida-betony-150-a-pound accessed 16/3/2021
14   Nilsen (2006)
15   Thayer (2006) pp.91-93

Chapter 12

1   annforfungi.co.uk
2   gallowaywildfoods.com/jelly-ear-fungus-edibility-identification-distribution
3   patents.google.com/patent/US4757640A/en

Books and articles

Barney, Paul (2016) 'Alliums for all seasons' at www.rareplantfair.co.uk/news-and-articles/alliums-for-all-seasons

Barceloux, Donald G. (2008) *Medical Toxicology of Natural Substances*

Barstow, Stephen (2014) *Around the World in 80 Plants*, Permanent Publications

Barstow, Stephen (2019) 'Spiked rampion: a perennial all season shade tolerant vegetable and one of the best edientomentals' at www.edimentals.com, Malvik, Norway

Betts, Edwin M., ed. (1944) *Thomas Jefferson's Garden Book, 1766-1824*: With Relevant Extracts from His Other Writings

Bradley, Martha (1760) *The British Housewife*, London: S. Crowder and H. Woodgate

Brill, Steve (2002, 2010) *The Wild Vegan Cookbook*, The Harvard Common Press

Brussell, David (2004) 'Araliaceae species used for culinary and medicinal purposes in Niigata-ken, Japan.' *Economic Botany* 58, pp.736-739

Bryant, Charles (1783) *Flora Dietetica: Or, History of Esculent Plants*

Coffey, Timothy (1993) *The History and Folklore of North American Wildflowers*, Facts on File

Cordero, S., Abello, L. and Galvez, F. (2017) P*lantas silvestres comestibles y medicinales de Chile y otras partes del mundo (Guía de Campo)* Ed. Corporación Chilena de la Madera, Concepción, Chile

Cox, Kenneth and Beaton, Caroline (2012) *Fruit and Vegetables for Scotland*, Birlinn Ltd.

Crawford, Martin (2010) *Creating a Forest Garden*, Green Books

Eaton, Mary (1822) *The Cook and Housekeeper's Dictionary*

Fern, Ken (1997) *Plants For A Future*, Permanent Publications

Ghirardini et al. (2007) 'The importance of a taste. A comparative study on wild food plant consumption in twenty-one local communities in Italy.' *Journal of Ethnobiology and Ethnomedicine* 3: 22

Gibbons, Euell (1962) *Stalking The Wild Asparagus*, David McKay, New York

Glasse, Hannah (1747) *The Art of Cookery, Made Plain and Easy*

Hart, Robert A. de J. (1991, 1996 revised) *Forest Gardening: Rediscovering Nature and Community in a Post-Industrial Age*, Green Books

Hashimoto, Ikuzo (2007) *Taberareru Yasei Shokubutsu Daijiten (Encyclopaedia of Edible Wild Plants)*, Kashiwa Shobo, Tokyo

Hendrick, U.P. (1919) *Sturtevant's Edible Plants of the World*, General Publishing Company, Toronto

Howell, M. (2019) 'The new 'superfood' you've never heard of – the berry with 15 times more Vitamin C than an orange', *Daily Telegraph*, 11 September 2019

Hu, Shiu-ying (2012) *Food Plants of China* (Vol. 1), Info Rainbow Ltd. for Chinese University of Hong Kong, Kindle Edition

Irving, Miles (2009) *The Forager Handbook: A Guide to the Edible Plants of Britain*, Ebury Press

Jacke, Dave and Toensmeier, Eric (2015) *Edible Forest Gardens*, Chelsea Green Publishing Co.

Kang et al. (2012) 'Wild food plants and wild edible fungi in two valleys of the Qinling Mountains (Shaanxi, central China).' *Journal of Ethnobiology and Ethnomedicine*, 2012 9:26

Kunkel, Günther (1984) *Plants for Human Consumption*, Koeltz Scientific Books

Ladio, Ana H. and Lozada, Mariana (2004) 'Patterns of use and knowledge of wild edible plants in distinct ecological environments: a case study of a Mapuche community from northwestern Patagonia.' *Biodiversity and Conservation* (2004) 13: 1153. p.253

Linnaeus, Carl (1737) *Flora Lapponica*

Martynoga, Fi, ed. (2012) *A Handbook of Scotland's Wild Harvests*, Saraband for Reforesting Scotland and Scottish Wild Harvests Association

Martins, S., Simões, F., Matos, J., Silva, A.P. and Carnide, V. (2014). 'Genetic relationship among wild, landraces, and cultivars of hazelnut (Corylus avellana) from Portugal revealed through ISSR and AFLP markers.' *Plant Systematics and Evolution*. 300 (5): 1035–1046.

Mac Coitir, Niall (2003, 2015) *Ireland's Wild Plants: Myths Legends and Folklore*, The Collins Press

McGee, Harold (2007) *On Food and Cooking: The Science and Lore of the Kitchen*, New York: Scribner

Marten, Gerald G. (1986) *Traditional Agriculture in Southeast Asia: A Human Ecology Perspective*, Westview Press, Boulder, Colorado

Mears, Ray and Hillman, Gordon (2007) *Wild Food*, Hodder & Stoughton

Mithen, Steven, Finlay, Nyree; Carruthers, Wendy, Carter, Stephen, and Ashmore, Patrick (2001) 'Plant Use in the Mesolithic: Evidence from Staosnaig, Isle of Colonsay, Scotland.' *Journal of Archaeological Science*. 28. 223-234. 10.1006/jasc.1999.0536.

Moerman, Daniel E. (2010) *Native American Plant Foods, An Ethnobotanical Dictionary*, Timber Press, Portland, Oregon

Nie et al. (2019) 'Giant Pandas Are Macronutritional Carnivores.' *Current Biology* 29, 1–6 May 20, 2019

Nilsen, Gerd (2006) 'Cloudberries—The Northern Gold.' *International Journal of Fruit Science* 5. 45-60. 10.1300/J492v05n02_06.

Phillips, Roger (1983) *Wild Food*, Peerage Books

Ryynänen, A. (1973) '*Rubus arcticus* L. and its cultivars.' *Annales Agriculturae Fennae* 12: 1-76

Smith, K. Annabelle (2013) 'Why the Tomato Was Feared in Europe for More Than 200 Years.' *Smithsonian Magazine*, June 18, 2013

Soemarwoto, Otto; Conway, Gordon R. (1992) 'The Javanese homegarden.' *Journal of Farming Systems Research-Extension*. 2 (3): 95–118.

Svanberg, Ingvar (2012) 'Gathering dog's tooth violet (*Erythronium sibiricum*) in Siberia.' *Journal de la Société Finno-Ougrienne* 93 (2012), pp.441–454

Thayer, Samuel (2006) *The Forager's Harvest*, Forager's Harvest Press

Thayer, Samuel (2010) *Nature's Garden: A Guide to Identifying, Harvesting, and Preparing Edible Wild Plants*, Forager's Harvest Press

Thayer, Samuel (2017) *Incredible Wild Edibles*, Forager's Harvest Press

Turner, Nancy J. (1995) *Food Plants of Coastal First Peoples*, Royal BC Museum

Turner, Nancy J. (1997) *Food Plants of Interior First Peoples*, Royal BC Museum

Wilde, Monica (2018) 'Food from the forest: common nettle.' *Reforesting Scotland Journal*, Issue 57

Yatskievych, George (2013) *Steyermark's Flora of Missouri, Volume 3*

Yıldırım, E., Dursun, A. and Turun, M. (2001) 'Determination of the Nutrition Contents of the Wild Plants Used as Vegetables in Upper Çoruh Valley.' *Turkish Journal of Botany* 25 (2001) 367-371

Blogs and websites

Ann Miller's Speciality Mushrooms,
    www.annforfungi.co.uk
Bburi Kitchen, bburikitchen.com
Botanic Garden, www.botanic.jp
Chileflora, www.chileflora.com
Cultivariable, www.cultivariable.com
Daylily BReeder [*sic*], daylilybreeder.blogspot.com
Eat the Weeds, www.eattheweeds.com
Eat Weeds, www.eatweeds.co.uk
Edible Manhattan, www.ediblemanhattan.com
Edible Wild Food, www.ediblewildfood.com
Edimentals, www.edimentals.com
Flora of China, www.efloras.org
Galloway Wild Foods, gallowaywildfoods.com
Garden Myths, www.gardenmyths.com
Guadua Bamboo, www.guaduabamboo.com/blog/
    edible-bamboo-species

Honest Food, honest-food.net
Lewis Bamboo, lewisbamboo.com/
    edible-bamboo-species
Maangchi, www.maangchi.com
Narrative Environments,
    www.narrativeenvironments.ch
Nettle Beer, nettlebeer.com
Online Atlas of the British and Irish Flora,
    www.brc.ac.uk/plantatlas
Permies, permies.com
Plants For A Future, pfaf.org
Scottish Bamboo, www.scottishbamboo.com
Shikigami, nakazora.wordpress.com
Shizuoka Gourmet, shizuokagourmet.com
Tree Guidebook by Kanon, kanon1001.web.fc2.com
Wild Food Plants of Britain,
    foragerplants.blogspot.com

Photo credits

279

Index

Solutions for Changemakers

Every one of us who picks up a seed packet, upcycles a piece of furniture or invests time and energy in a community project is a changemaker. The minute we start to make small but incremental changes in our lives, we become agents for positive change and the ripples of our influence spread out around us.

Permanent Publications produce a range of books to empower and inspire changemakers the world over, from no dig organic growing, food forests and permaculture design, to natural building, renewable technology and connecting with Nature.

If you enjoyed *A Food Forest in Your Garden*, why not try these titles related to forest gardening and organic gardening:

SHRUBS FOR GARDENS AGROFORESTRY AND PERMACULTURE

Martin Crawford

£24.00

THE PLANT LOVER'S BACKYARD FOREST GARDEN

Pippa Chapman

£24.95

FOREST GARDENING IN PRACTICE

Tomas Remiarz

£16.00

For the full list of our solution-based titles visit
www.permanentpublications.co.uk

Our books are also available via our American distributor, Chelsea Green at
www.chelseagreen.com/publisher/permanent-publications

Permanent Publications also publishes *Permaculture* magazine

Join the Changemakers

We also publish *Permaculture* magazine,
a voice for changemakers worldwide.

Permaculture magazine gives you the tools to
create productive and resilient homes, gardens, economies,
relationships, schools, farms and communities.

Each issue of this visionary magazine offers tried-and-tested
solutions, projects and pioneering ideas from the very best of
the permaculture community.

And all subscriptions, print or digital, come with free digital
access to our archive – that's over 30 years of the permaculture
world to immerse in: **www.permaculture.co.uk/subscribe**

Join our family

▶ permaculturemedia

📷 permaculture
magazine

f PermacultureMag

🐦 PermacultureMag

www.permaculture.co.uk